Introduction

Welcome to "Curiosity Chronicles: Astonishing Facts for Inquisitive Minds." This book is your gateway to an extraordinary journey through the realms of science, technology, and the marvels of the human body. Whether you're a curious mind, a science enthusiast, or someone who simply loves learning new things, you're about to embark on a thrilling adventure filled with fascinating facts that will amaze, inform, and inspire you.

Interactive and Engaging

Fun Trivia: Test your knowledge with engaging trivia questions scattered throughout the book, sparking conversations and challenging your understanding of these topics.

In-Depth Explanations: Each fact is accompanied by a brief explanation to provide context and enhance your learning experience.

Plus, enjoy stunning images that bring the facts to life, making your reading experience even more engaging!

What to Expect Inside

- **Space:**
  - Explore the Universe: Dive into the mysteries of the cosmos, from the formation of stars and black holes to the vastness of galaxies and the wonders of our solar system.
- **Earth:**
  - Discover Our Planet: Uncover the secrets of Earth's geological marvels, diverse ecosystems, and awe-inspiring natural phenomena.
  - Natural Wonders: From the highest peaks to the deepest oceans, explore the fascinating features that make our planet unique.
- **Animals:**
  - Animal Kingdom: Delve into the behaviors, adaptations, and diversity of the animal world and learn about the incredible capabilities and unique traits of various animal species.
- **Human Body:**
  - Intricacies of the Human Body: Explore the skeletal, muscular, nervous, circulatory, respiratory, digestive, sensory, immune, and endocrine systems.
  - Human Marvels: Discover astonishing facts about the human body's abilities, quirks, and lesser-known features that make us extraordinary.
- **Technology:**
  - Technological Evolution: Trace the history of technology from groundbreaking milestones to modern gadgets, computing, and the internet.
  - Innovations and Trends: Dive into the world of artificial intelligence, robotics, cybersecurity, and the cutting-edge technologies shaping our future.

Fun Fact

If you read one fact per minute, it would take you 10 hours to finish this book. That's almost as long as it takes to binge-watch an entire season of your favorite TV show!

Welcome to the Wonders of Space!

Welcome to the exciting world of space! In this section, we'll explore amazing facts about our universe, from planets and stars to black holes and space travel. Get ready to embark on a journey through the cosmos!

- **Fact 1**: There are more stars in the universe than grains of sand on all the Earth's beaches.

    "The universe is incredibly vast and contains an estimated 1 septillion stars (1 followed by 24 zeros). This mind-boggling number far exceeds the number of grains of sand on Earth, showcasing the enormity of space."

- **Fact 2**: One day on Venus is longer than a year on Venus.

  "Venus has an extremely slow rotation on its axis, taking about 243 Earth days to complete one rotation. In contrast, it takes about 225 Earth days for Venus to orbit the Sun. This means a single day on Venus lasts longer than its entire year."

- **Fact 3**: A teaspoon of a neutron star would weigh about 6 billion tons.

  "Neutron stars are incredibly dense remnants of supernova explosions. Their gravity compresses their mass into a small volume, resulting in immense density. A small amount of neutron star material would be incredibly heavy due to this density."

- **Fact 4**: The largest volcano in our solar system is on Mars.

  "Olympus Mons, located on Mars, is the largest volcano in the solar system. It stands about 13.6 miles (22 kilometers) high and is approximately 370 miles (600 kilometers) in diameter, making it nearly three times the height of Mount Everest."

- **Fact 5**: Black holes are invisible because light cannot escape them.

  "Black holes have such strong gravitational pulls that not even light can escape once it passes the event horizon. This makes black holes invisible to the naked eye, though their presence can be inferred by their effects on nearby objects and light."

- **Fact 6**: The first manned moon landing was in 1969.

    "On July 20, 1969, NASA's Apollo 11 mission successfully landed astronauts Neil Armstrong and Buzz Aldrin on the Moon. Armstrong's famous words, "That's one small step for man, one giant leap for mankind," marked this historic event."

- **Fact 7**: The Milky Way galaxy is on a collision course with the Andromeda galaxy.

    "Our Milky Way galaxy is slowly moving towards the Andromeda galaxy. Scientists predict that in about 4 billion years, the two galaxies will collide and eventually merge to form a new galaxy, often referred to as "Milkomeda" or "Milkdromeda.""

- **Fact 8**: The hottest planet in our solar system is Venus.

    "Despite being second from the Sun, Venus is the hottest planet due to its thick atmosphere, which traps heat in a runaway greenhouse effect. Surface temperatures on Venus can reach up to 900°F (475°C)."

- **Fact 9**: The International Space Station (ISS) orbits Earth about 16 times a day.

    "The ISS travels at a speed of about 17,500 miles per hour (28,000 kilometers per hour), allowing it to orbit Earth approximately 16 times each day. This rapid orbit results in the crew experiencing 16 sunrises and sunsets every 24 hours."

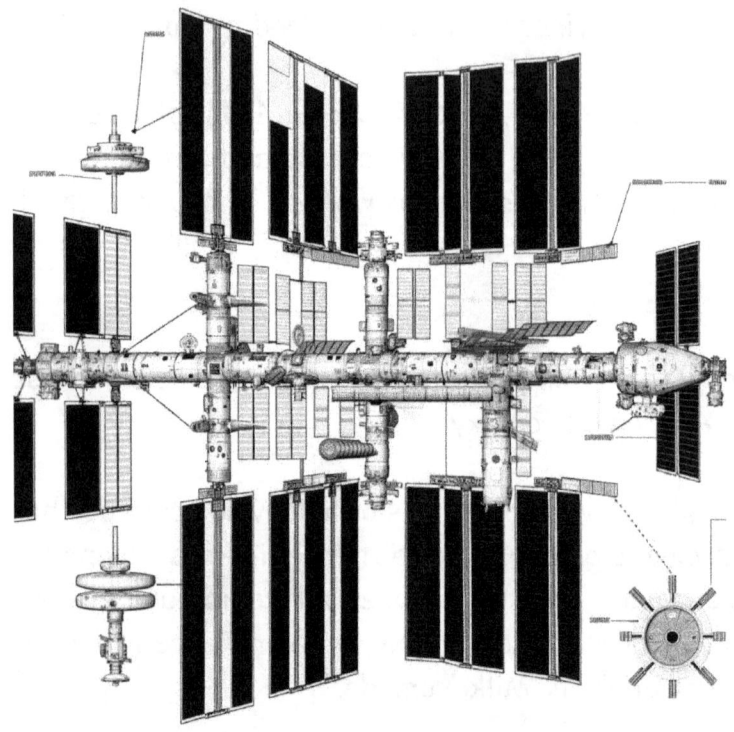

- **Fact 10**: The Sun makes up about 99.86% of the mass in the solar system.

    "The Sun, being the largest and most massive object in our solar system, contains nearly all of its mass. This immense mass exerts a gravitational force that keeps the planets, moons, and other objects in their orbits around the Sun."

- **Fact 11**: There are about 100 billion galaxies in the observable universe.

    "The observable universe is a vast expanse containing an estimated 100 billion galaxies, each with millions or even billions of stars. This estimate continues to grow as telescopes improve and our understanding of the universe expands."

- **Fact 12**: The Great Red Spot on Jupiter is a giant storm.

    "The Great Red Spot is a massive storm on Jupiter that has been raging for at least 400 years. It is about 1.3 times the diameter of Earth and is characterized by high-speed winds and turbulent clouds."

- **Fact 13**: Saturn's rings are made of ice and rock.

    "Saturn's iconic rings consist of countless small particles, primarily composed of water ice, with traces of rock and dust. These particles range in size from tiny grains to large chunks several meters across."

- **Fact 14**: The Moon is moving away from Earth at a rate of about 1.5 inches per year.

    "Due to gravitational interactions, the Moon is slowly drifting away from Earth at a rate of approximately 1.5 inches (3.8 centimeters) per year. This gradual separation has been occurring for billions of years."

- **Fact 15**: The Hubble Space Telescope has provided images of galaxies over 13 billion years old

    "Launched in 1990, the Hubble Space Telescope has captured detailed images of distant galaxies, some dating back to over 13 billion years. These images offer a glimpse into the early universe and its formation."

- **Fact 16**: A year on Neptune is equivalent to 165 Earth years.

    "Neptune takes about 165 Earth years to complete one orbit around the Sun. This long orbital period means that a single year on Neptune lasts more than a century and a half by Earth standards."

- **Fact 17**: The Sun's core temperature is about 27 million degrees Fahrenheit.

    "The core of the Sun reaches temperatures of approximately 27 million degrees Fahrenheit (15 million degrees Celsius). This extreme heat results from nuclear fusion, where hydrogen atoms combine to form helium, releasing vast amounts of energy."

- **Fact 18**: The Kuiper Belt is a region of the solar system beyond Neptune.

    "The Kuiper Belt is a distant region of the solar system extending beyond Neptune's orbit. It contains many small icy bodies, including dwarf planets like Pluto, and is believed to be a remnant of the early solar system."

- **Fact 19**: The speed of light is about 186,282 miles per second.

   "Light travels at an astonishing speed of approximately 186,282 miles per second (299,792 kilometers per second). This speed limit of the universe means that light from the Sun takes about 8 minutes to reach Earth."

- **Fact 20**: Mars has the largest canyon in the solar system.

   "Valles Marineris on Mars is the largest canyon in the solar system, stretching over 2,500 miles (4,000 kilometers) long and reaching depths of up to 7 miles (11 kilometers). This immense canyon dwarfs Earth's Grand Canyon."

- **Fact 21**: The temperature on the Moon can range from -298°F to 224°F.

   "The Moon experiences extreme temperature fluctuations due to its lack of atmosphere. During the lunar day, temperatures can soar to 224°F (107°C), while at night, they can plummet to -298°F (-183°C)."

- **Fact 22**: The largest known star is UY Scuti.

   "UY Scuti, a red supergiant star, is the largest known star by radius. It is approximately 1,700 times the size of the Sun, with a volume that could contain billions of Suns."

- **Fact 23**: The Oort Cloud is a theoretical cloud of icy bodies surrounding the solar system.

  "The Oort Cloud is a hypothetical region thought to contain countless icy objects. It is believed to be the source of long-period comets and extends far beyond the Kuiper Belt, encircling the solar system."

- **Fact 24**: The nearest star to Earth, after the Sun, is Proxima Centauri.

  "Proxima Centauri, part of the Alpha Centauri star system, is the closest star to Earth beyond the Sun. It is located about 4.24 light-years away, making it a potential target for future interstellar missions."

- **Fact 25**: The surface of Mars is covered in iron oxide, giving it a reddish color.

  "Mars is often called the "Red Planet" due to its surface being coated in iron oxide, or rust. This reddish hue is a defining characteristic of the planet and can be seen from Earth with the naked eye."

- **Fact 26**: The Andromeda Galaxy is the closest spiral galaxy to the Milky Way.

  "The Andromeda Galaxy, located about 2.537 million light-years from Earth, is the nearest spiral galaxy to our Milky Way. It is expected to collide and merge with our galaxy in about 4 billion years."

- **Fact 27**: A light-year is the distance light travels in one year.

  "A light-year measures the distance that light travels in one year, which is about 5.88 trillion miles (9.46 trillion kilometers). This unit helps astronomers express vast distances in space."

- **Fact 28**: There are regions of space where time itself slows down due to gravity.

  "According to Einstein's theory of general relativity, strong gravitational fields, like those near black holes, can cause time to slow down. This phenomenon, known as time dilation, means that time passes more slowly in intense gravitational fields."

- **Fact 29**: Some of the most powerful telescopes are located in space.

  "Space telescopes, like the Hubble Space Telescope and the upcoming James Webb Space Telescope, are placed in orbit to avoid atmospheric interference, allowing them to capture clearer and more detailed images of the universe."

- **Fact 30**: Saturn has 82 known moons.

    "Saturn is surrounded by a diverse collection of 82 confirmed moons, with Titan being the largest. These moons vary in size and composition, contributing to the complexity and beauty of the Saturnian system."

- **Fact 31**: Pluto was reclassified as a dwarf planet in 2006.

    "In 2006, the International Astronomical Union (IAU) redefined the criteria for planetary status, leading to Pluto's reclassification as a dwarf planet. This decision was based on Pluto's size, orbit, and inability to clear its orbital path of other debris."

- **Fact 32**: The Sun's atmosphere, or corona, is hotter than its surface.

    "The Sun's corona, its outermost layer, has temperatures reaching millions of degrees Fahrenheit, far hotter than the Sun's surface, which is about 10,000°F. The exact reason for this temperature difference remains a topic of scientific investigation."

- **Fact 33**: Uranus rotates on its side.

    "Unlike other planets, Uranus has an extreme axial tilt of about 98 degrees, causing it to rotate on its side. This unique tilt likely resulted from a collision with a large object during the planet's formation."

- **Fact 34**: The largest moon of Neptune is Triton.

    "Triton, Neptune's largest moon, is unique due to its retrograde orbit, meaning it orbits Neptune in the opposite direction of the planet's rotation. This suggests Triton may have been captured by Neptune's gravity rather than forming in place."

- **Fact 35**: The speed needed to escape Earth's gravity is about 25,000 miles per hour.

    "To break free from Earth's gravitational pull, an object must reach a velocity of approximately 25,000 miles per hour (40,270 kilometers per hour), known as the escape velocity. Rockets must achieve this speed to enter space."

- **Fact 36**: The Sun contains 99.86% of the mass in the solar system.

    "The Sun's immense mass accounts for nearly all the mass in the solar system, exerting a strong gravitational pull that keeps the planets, moons, and other objects in orbit around it."

- **Fact 37**: Mercury has no atmosphere to retain heat, causing extreme temperature variations.

    "Without an atmosphere, Mercury experiences drastic temperature changes, ranging from -330°F at night to 800°F during the day. This lack of atmospheric insulation leads to its severe temperature extremes."

- **Fact 38**: The largest asteroid in the asteroid belt is Ceres.

  "Ceres, the largest object in the asteroid belt between Mars and Jupiter, is classified as a dwarf planet. It is about 590 miles (940 kilometers) in diameter and was the first asteroid discovered in 1801."

- **Fact 39**: The farthest spacecraft from Earth is Voyager 1.

  "Launched in 1977, Voyager 1 is the most distant human-made object from Earth. It has traveled beyond our solar system into interstellar space, sending back valuable data about the outer reaches of our solar system and beyond."

- **Fact 40**: Space is completely silent.

  "In space, there is no air or other medium for sound waves to travel through, making it a completely silent environment. Astronauts communicate via radio waves, which can travel through the vacuum of space."

- **Fact 41**: The Sun will eventually become a red giant.

  "In about 5 billion years, the Sun will exhaust its hydrogen fuel and expand into a red giant, engulfing the inner planets, including Earth. It will then shed its outer layers, leaving behind a white dwarf."

- **Fact 42**: There is water ice on the Moon.

  "Scientists have confirmed the presence of water ice in permanently shadowed craters at the Moon's poles.

- **Fact 43**: The hottest place in the universe is the core of a star.

  "The cores of stars, where nuclear fusion occurs, reach temperatures of millions of degrees Fahrenheit. This extreme heat is necessary for the fusion process that powers stars and produces energy."

- **Fact 44**: Saturn's moon Titan has lakes of liquid methane.

  "Titan, the largest moon of Saturn, has a thick atmosphere and surface lakes and rivers of liquid methane and ethane. This makes it one of the most Earth-like bodies in the solar system, albeit with very different chemistry."

- **Fact 45**: A day on Mars is just over 24 hours.

  "Mars has a rotation period of 24.6 hours, making a Martian day, or sol, only slightly longer than an Earth day. This similarity in day length is one factor that makes Mars an interesting target for human exploration."

- **Fact 46**: The largest impact crater in the solar system is on Mars.

  "The Hellas Basin on Mars is the largest known impact crater which measures about 1,400 miles (2,300 kilometers) in diameter and is over 5 miles (8 kilometers) deep."

- **Fact 47**: Jupiter has the strongest magnetic field of any planet in the solar system.

  "Jupiter's magnetic field is about 20,000 times stronger than Earth's. This powerful field creates intense radiation belts and influences the planet's many moons."

- **Fact 48**: The average distance between the Earth and the Sun is about 93 million miles.

  "This distance, known as an astronomical unit (AU), is used as a standard measure for distances within our solar system. It takes light from the Sun about 8 minutes to reach Earth."

- **Fact 49**: There are more than 100 moons in the solar system.

  "The planets and dwarf planets in our solar system have over 100 moons collectively. These moons vary in size, composition, and geological activity, with some showing signs of subsurface oceans and potential for life."

- **Fact 50**: The Sun's gravity keeps the planets in orbit.

  "The Sun's immense gravitational pull ensures that the planets, along with other objects like asteroids and comets, remain in their orbits around it. This gravitational force is a fundamental aspect of the solar system's stability."

- **Fact 51**: Mars has the largest dust storms in the solar system.

    "Dust storms on Mars can cover the entire planet and last for weeks or even months. These storms are driven by the planet's thin atmosphere and can significantly impact surface conditions."

- **Fact 52**: The first artificial satellite, Sputnik 1, was launched by the Soviet Union in 1957.

    "Sputnik 1 marked the beginning of the space age and the start of space exploration. This historic launch paved the way for future missions and technological advancements in space."

- **Fact 53**: The Milky Way is a barred spiral galaxy.

    "Our galaxy, the Milky Way, has a central bar-shaped structure made of stars, with spiral arms extending from it. This structure influences the motion and distribution of stars and other matter within the galaxy."

- **Fact 54**: Neptune was discovered using mathematical predictions.

    "Astronomers predicted Neptune's existence based on irregularities in Uranus's orbit. Johann Galle and Heinrich d'Arrest confirmed its position in 1846, making it the first planet discovered through mathematical calculations."

- **Fact 55**: A comet's tail always points away from the Sun.

    "As a comet approaches the Sun, solar radiation and the solar wind cause the comet's tail to form and point away from the Sun. This occurs regardless of the comet's direction of travel."

- **Fact 56**: The Moon's gravity affects tides on Earth.

    "The gravitational pull of the Moon causes the Earth's oceans to bulge, creating high and low tides. This effect is most noticeable in coastal areas and is a fundamental aspect of our planet's interaction with its satellite."

- **Fact 57**: Venus has more volcanoes than any other planet in the solar system.

    "Venus is covered with thousands of volcanoes, ranging from small lava domes to massive shield volcanoes. Volcanic activity has played a significant role in shaping the planet's surface."

- **Fact 58**: The closest planet to the Sun, Mercury, has a very thin atmosphere.

    "Mercury's thin atmosphere, known as an exosphere, is composed mainly of oxygen, sodium, and hydrogen. This sparse atmosphere offers little protection from the Sun's radiation and solar wind."

- **Fact 59**: The first human-made object to reach the Moon was the Soviet Luna 2 in 1959.

    "Luna 2, an unmanned spacecraft, was the first human-made object to impact the Moon. Its successful mission provided valuable data about the Moon's surface and paved the way for future lunar exploration."

- **Fact 60**: The temperature in space is about -455°F.

    "Space is incredibly cold, with a temperature of about -455°F (-270°C) in the vast regions between stars and galaxies. This near-absolute-zero temperature is due to the lack of atmosphere and the minimal presence of particles."

- **Fact 61**: The Great Wall of China is not visible from space without aid.

    "Despite popular belief, the Great Wall of China is not visible to the naked eye from space. Astronauts can see the wall using binoculars or cameras, but it blends with the natural landscape from a distance."

- **Fact 62**: The first American in space was Alan Shepard.

  "Alan Shepard became the first American to travel into space on May 5, 1961, aboard the Mercury-Redstone 3 mission, also known as Freedom 7. His suborbital flight lasted about 15 minutes."

- **Fact 63**: The Sun's magnetic field reverses every 11 years.

  "The Sun undergoes a solar cycle approximately every 11 years, during which its magnetic field flips. This cycle includes periods of high and low solar activity, impacting space weather and solar phenomena."

- **Fact 64**: Saturn's moon Enceladus has geysers that eject water into space.

  "Enceladus, one of Saturn's moons, has active geysers near its south pole that shoot water vapor, ice particles, and organic molecules into space. These geysers suggest the presence of a subsurface ocean."

- **Fact 65**: The speed of sound is slower on Mars than on Earth.

    "The speed of sound on Mars is slower than on Earth due to the thin Martian atmosphere, composed mostly of carbon dioxide. Sound travels at about 537 mph (864 km/h) on Mars compared to 767 mph (1,235 km/h) on Earth."

- **Fact 66**: Jupiter's moon Ganymede is the largest moon in the solar system.

    "Ganymede, a moon of Jupiter, is the largest moon in the solar system, with a diameter of about 3,273 miles (5,268 kilometers). It is even larger than the planet Mercury."

- **Fact 67**: The Earth is not a perfect sphere.

    "Earth is an oblate spheroid, meaning it is slightly flattened at the poles and bulges at the equator due to its rotation. This shape results from the centrifugal force caused by Earth's spin."

- **Fact 68**: The first spacecraft to leave the solar system was Voyager 1.

    "Voyager 1, launched by NASA in 1977, became the first spacecraft to enter interstellar space in 2012. It continues to send back data about the outer regions of our solar system."

- **Fact 69**: The Moon has quakes called "moonquakes."

    "Moonquakes are seismic activities on the Moon, caused by tidal forces from Earth and thermal expansion. These quakes are less intense than earthquakes but can last for extended periods due to the Moon's rigid structure."

- **Fact 70**: There are active volcanoes on Jupiter's moon Io.

    "Io, one of Jupiter's moons, is the most volcanically active body in the solar system. Its surface is dotted with active volcanoes that frequently erupt, driven by intense tidal forces from Jupiter."

- **Fact 71**: The Kuiper Belt is home to many dwarf planets.

    "The Kuiper Belt, a region beyond Neptune, contains numerous small icy bodies, including dwarf planets like Pluto, Haumea, and Makemake. This region is a remnant of the early solar system."

- **Fact 72**: The Sun is about halfway through its life cycle.

    "The Sun, currently about 4.6 billion years old, is in the middle of its main sequence phase, where it burns hydrogen into helium. It is expected to continue for another 5 billion years before evolving into a red giant."

- **Fact 73**: Neptune has supersonic winds.

    "Neptune experiences some of the fastest winds in the solar system, reaching speeds of up to 1,200 miles per hour (1,931 kilometers per hour). These supersonic winds are driven by the planet's internal heat and dynamic atmosphere."

- **Fact 74**: Mercury has water ice in permanently shadowed craters.

    "Despite its proximity to the Sun, Mercury has water ice in permanently shadowed craters at its poles. These regions never receive sunlight, allowing ice to persist in the extreme cold."

- **Fact 75**: The Sun rotates faster at its equator than at its poles.

    "The Sun exhibits differential rotation, where its equator rotates faster than its poles. The equator completes a rotation in about 25 days, while the poles take around 35 days."

- **Fact 76**: A day on Neptune lasts about 16 hours.

  "Neptune has a rapid rotation period, with a single day lasting about 16 hours. This fast rotation contributes to the planet's dynamic weather and intense wind patterns."

- **Fact 77**: The Solar System is located in the Milky Way's Orion Arm.

  "Our Solar System resides in a minor spiral arm of the Milky Way galaxy called the Orion Arm, or Orion Spur. This region is situated between the larger Perseus and Sagittarius arms."

- **Fact 78**: The first woman in space was Valentina Tereshkova.

  "Valentina Tereshkova, a Soviet cosmonaut, became the first woman to travel into space on June 16, 1963, aboard Vostok 6. Her mission lasted almost three days."

- **Fact 79**: There are black holes with masses billions of times that of the Sun.

  "Supermassive black holes, located at the centers of galaxies, can have masses billions of times greater than the Sun. These enormous black holes play a crucial role in the formation and evolution of galaxies."

- **Fact 80**: The first successful Mars rover was Sojourner.

  "NASA's Sojourner rover, part of the Mars Pathfinder mission, successfully landed on Mars on July 4, 1997, and conducted experiments on the Martian surface for nearly three months."

- **Fact 84**: There are rogue planets that do not orbit a star.

    "Rogue planets, also known as free-floating planets, drift through space without orbiting any star. These planets are thought to have been ejected from their original star systems."

- **Fact 85**: The first soft landing on the Moon was by Luna 9.

    "The Soviet Luna 9 mission achieved the first soft landing on the Moon on February 3, 1966, sending back the first close-up images of the lunar surface."

- **Fact 86**: The Sun's surface is called the photosphere.

    "The photosphere is the visible surface layer of the Sun, emitting the light we see from Earth. It has an average temperature of about 10,000°F (5,500°C)."

- **Fact 87**: Space probes have visited every planet in the solar system.

    "NASA and other space agencies have successfully sent probes to explore all the planets in our solar system, providing valuable data and images of these distant worlds."

- **Fact 88**: The Sun's solar wind extends beyond Pluto.

    "The solar wind, a stream of charged particles released by the Sun, extends well beyond Pluto, creating a bubble in space known as the heliosphere that protects the solar system from interstellar radiation."

- **Fact 89**: Saturn has the most extensive rings in the solar system.

  "Saturn's rings are made up of ice and rock particles ranging from tiny grains to large boulders, and they extend over 175,000 miles from the planet."

- **Fact 90**: There is a massive hexagonal storm at Saturn's north pole.

  "This persistent storm system, first observed by the Voyager missions, has a unique hexagonal shape and spans about 20,000 miles across."

- **Fact 91**: The observable universe is about 93 billion light-years in diameter.

  "This vast expanse contains all the galaxies, stars, and cosmic structures that we can observe from Earth."

- **Fact 92**: There is a planet made entirely of diamonds.

  "Scientists believe that the exoplanet 55 Cancri e, located about 40 light-years away, is composed largely of carbon in the form of diamond."

- **Fact 93**: A supernova is an exploding star.

  "Supernovae occur when massive stars exhaust their nuclear fuel and collapse, resulting in a cataclysmic explosion."

- **Fact 94**: The James Webb Space Telescope will be able to see the first galaxies formed.

    "This telescope, set to launch in 2021, will observe galaxies that formed shortly after the Big Bang, providing insights into the early universe."

- **Fact 95**: Mars has two moons, Phobos and Deimos.

    "These small, irregularly shaped moons are believed to be captured asteroids from the asteroid belt."

- **Fact 96**: The Great Attractor is a gravitational anomaly in intergalactic space.

    "It appears to be drawing galaxies, including our own Milky Way, toward a region in space at over 600 kilometers per second."

- **Fact 97**: The Horsehead Nebula is a dark nebula in the constellation Orion.

    "This iconic nebula gets its name from its horsehead-like shape and is part of a large, dark cloud of dust and gas."

- **Fact 98**: The Kepler Space Telescope discovered over 2,600 exoplanets.

    "Launched in 2009, Kepler significantly expanded our understanding of planets beyond our solar system."

- **Fact 99**: The first photograph of a black hole was taken in 2019.

    "Using the Event Horizon Telescope, scientists captured an image of the black hole at the center of the galaxy M87."

    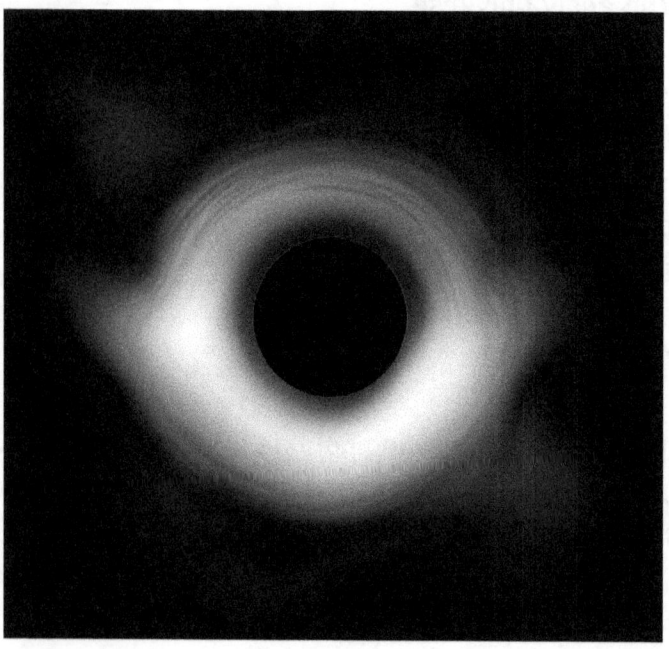

- **Fact 100**: The cosmic microwave background radiation is the afterglow of the Big Bang.

    "This faint radiation is the remnant heat from the Big Bang, providing evidence for the origin of the universe."

- **Fact 101**: The Pillars of Creation are part of the Eagle Nebula.

    "These towering columns of gas and dust are sites of active star formation, captured in stunning detail by the Hubble Space Telescope."

- **Fact 102**: The universe is expanding at an accelerating rate.

  "Observations of distant galaxies show that the expansion of the universe is speeding up, driven by a mysterious force called dark energy."

- **Fact 103**: The cosmic web is the large-scale structure of the universe.

  "Galaxies are arranged in a vast network of filaments and voids, resembling a cosmic web."

- **Fact 104**: Earth orbits the Sun in an elliptical path while our solar system orbits the center of the Milky Way galaxy.

  "Earth follows an elliptical orbit around the Sun, taking approximately 365.25 days to complete one revolution. Simultaneously, our entire solar system orbits the center of the Milky Way galaxy, which takes about 225-250 million years to complete one full orbit, a journey known as a cosmic year or galactic year."

Space Trivia Questions-1
(Answers at the end)

Question 1: What is the largest planet in our solar system?

A) Earth
B) Mars
C) Jupiter
D) Saturn

Question 2: Which planet is known as the "Red Planet"?

A) Venus
B) Mars
C) Mercury
D) Jupiter

Question 3: What is the closest star to Earth?

A) Proxima Centauri
B) Alpha Centauri
C) Betelgeuse
D) Sirius

Question 4: Which planet has the most moons?

A) Earth
B) Mars
C) Jupiter
D) Saturn

Question 5: What is the name of the first artificial satellite launched into space?
A) Explorer 1
B) Sputnik 1
C) Apollo 11
D) Voyager 1

Question 6: What is the largest volcano in the solar system?
A) Mauna Loa
B) Mount Everest
C) Olympus Mons
D) Vesuvius

Question 7: How long does it take for light from the Sun to reach Earth?
A) 1 minute
B) 8 minutes
C) 15 minutes
D) 30 minutes

Question 8: Which planet is known for its extensive ring system?

A) Neptune
B) Uranus
C) Jupiter
D) Saturn

Space Trivia Questions-2
(Answers at the end)

Question 9: What is the name of the first human to journey into outer space?

A) Neil Armstrong
B) Buzz Aldrin
C) Yuri Gagarin
D) John Glenn

Question 10: Which planet is the hottest in our solar system?

A) Mercury
B) Venus
C) Earth
D) Mars

Question 11: How many planets are in our solar system?

A) 7
B) 8
C) 9
D) 10

Question 12: Which space telescope has provided images of galaxies over 13 billion years old?

A) Hubble Space Telescope
B) James Webb Space Telescope
C) Chandra X-ray Observatory
D) Spitzer Space Telescope

Question 13: What is the most volcanically active body in the solar system?

A) Earth
B) Io
C) Venus
D) Mars

Question 14: Which planet is known for having a Great Red Spot?

A) Mars
B) Jupiter
C) Saturn
D) Neptune

Space Trivia Questions-3
(Answers at the end)

Question 16: How many Earth days does it take for Venus to complete one rotation on its axis?
A) 24
B) 58
C) 117
D) 243

Question 17: Which planet has the fastest winds in the solar system?
A) Earth
B) Mars
C) Neptune
D) Jupiter

Question 18: What is the name of the largest moon of Jupiter?
A) Titan
B) Ganymede
C) Callisto
D) Europa

Question 19: The Sun accounts for what percentage of the mass in the solar system?
A) 50%
B) 75%
C) 90%
D) 99.86%

Question 20: Which mission was the first to successfully land a rover on Mars?
A) Viking 1
B) Pathfinder
C) Spirit
D) Opportunity

Introduction to Earth

Welcome to the fascinating world of Earth! Our home planet is a dynamic and complex system, full of wonders and mysteries. In this section, we will explore the layers of the Earth, the power of volcanoes and earthquakes, the vastness of our oceans and the water cycle, the intricacies of weather and climate, and the breathtaking natural wonders that dot our planet. Get ready to discover some amazing facts about the Earth!

- **Fact 1**: The Earth has four main layers.

    "The Earth is composed of the crust, mantle, outer core, and inner core, each with distinct properties and compositions."

- **Fact 2**: The crust is the thinnest layer of the Earth.

    "The Earth's crust ranges from about 5 to 70 kilometers thick, depending on whether it's oceanic or continental."

- **Fact 3**: The mantle makes up about 84% of Earth's volume.

    "The mantle, composed mainly of silicate minerals, extends from the base of the crust to a depth of about 2,900 kilometers."

- **Fact 4**: The outer core is made of liquid iron and nickel.

    "The outer core, which lies beneath the mantle, is responsible for generating Earth's magnetic field."

- **Fact 5**: The inner core is as hot as the surface of the Sun.

    "Despite being solid due to immense pressure, the inner core's temperature reaches up to 9,932°F (5,500°C)."

- **Fact 6**: The lithosphere includes the crust and the uppermost mantle.

    "The lithosphere is divided into tectonic plates that float on the semi-fluid asthenosphere below."

- **Fact 7**: The asthenosphere is a zone of partially melted rock..

    "This layer, located below the lithosphere, allows tectonic plates to move."

- **Fact 8**: Earth's crust is composed mostly of oxygen, silicon, and aluminum.

  "These elements make up the majority of the Earth's crust by weight."

- **Fact 9**: The continental crust is older than the oceanic crust.

  "Continental crust can be billions of years old, while oceanic crust is usually less than 200 million years old."

- **Fact 10**: The Mohorovičić discontinuity marks the boundary between the crust and the mantle.

  "This boundary, also known as the Moho, is characterized by a sudden change in seismic wave speeds."

- **Fact 11**: There are about 1,500 potentially active volcanoes worldwide.

  "These volcanoes have erupted in the past 10,000 years and may erupt again."

- **Fact 12**: The Ring of Fire is home to 75% of the world's active and dormant volcanoes.

  "This horseshoe-shaped zone in the Pacific Ocean is known for frequent volcanic activity and earthquakes."

- **Fact 13**: The largest volcanic eruption recorded in history occurred at Mount Tambora in 1815.

    "This eruption in Indonesia released massive amounts of ash and gas, leading to the "Year Without a Summer."

- **Fact 14**: Hawaii was formed by volcanic activity.

    "The Hawaiian Islands are the result of a hotspot beneath the Pacific Plate, which creates new volcanic islands over millions of years."

- **Fact 15**: An earthquake's magnitude is measured using the Richter scale.

    "The Richter scale quantifies the amount of energy released during an earthquake, with each whole number representing a tenfold increase in amplitude."

- **Fact 16**: The San Andreas Fault is a major source of earthquakes in California.

  "This transform fault marks the boundary between the Pacific and North American tectonic plates."

- **Fact 17**: Tsunamis are often caused by underwater earthquakes.

  "These powerful waves are triggered by the displacement of large volumes of water due to seismic activity."

- **Fact 18**: Lava is molten rock that erupts onto the Earth's surface.

  "When magma reaches the surface, it is called lava and can flow and solidify, forming new land."

- **Fact 19**: The largest shield volcano in the solar system is Olympus Mons on Mars.

  "Although not on Earth, this comparison highlights the scale of volcanic features."

- **Fact 20**: Hotspots can create volcanic island chains.

  "As a tectonic plate moves over a stationary hotspot, a series of volcanoes can form, creating island chains like Hawaii."

- **Fact 21**: Oceans cover more than 70% of the Earth's surface.

    "The Earth's oceans play a crucial role in regulating climate and supporting marine life."

- **Fact 22**: The Pacific Ocean is the largest and deepest ocean.

    "Covering more than 63 million square miles, the Pacific Ocean contains the deepest point on Earth, the Mariana Trench."

- **Fact 23**: The water cycle is the continuous movement of water on, above, and below the Earth's surface.

    "This cycle involves processes such as evaporation, condensation, precipitation, and infiltration."

- **Fact 24**: The Gulf Stream is a powerful ocean current that influences the climate of the east coast of North America and Europe.

    "This warm current flows from the Gulf of Mexico up the eastern seaboard and across the Atlantic Ocean."

- **Fact 25**: The Great Barrier Reef is the largest coral reef system in the world.

    "Located off the coast of Australia, it spans over 1,400 miles and is home to a diverse range of marine species."

- **Fact 26**: The Arctic Ocean is the smallest and shallowest ocean.

    "Despite its size, the Arctic Ocean plays a vital role in Earth's climate and supports unique ecosystems."

- **Fact 27**: The Dead Sea is one of the saltiest bodies of water on Earth.

    "Located between Jordan and Israel, the Dead Sea has a salinity of about 34%, making it impossible for most life forms to survive."

- **Fact 28**: Water makes up about 60% of the human body.

    "This essential component is vital for various bodily functions, including temperature regulation and waste removal."

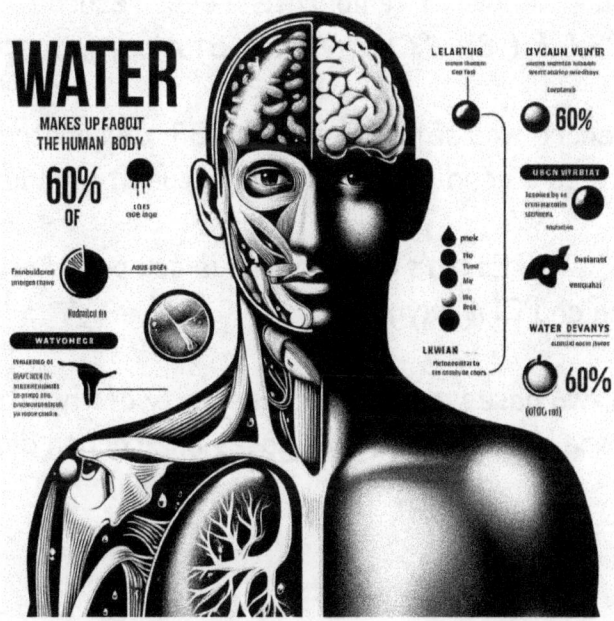

- **Fact 29**: The Amazon River is the largest river by discharge volume.

    "Flowing through South America, the Amazon River discharges more water than any other river, significantly impacting the global water cycle."

- **Fact 30**: Icebergs are formed from freshwater ice.

    "Icebergs calve from glaciers and ice sheets, floating in the ocean and slowly melting over time."

- **Fact 31**: The highest temperature ever recorded on Earth was 134°F (56.7°C) in Death Valley, California.

    "This record was set on July 10, 1913, highlighting the extreme heat conditions that can occur on our planet."

- **Fact 32**: The lowest temperature ever recorded on Earth was -128.6°F (-89.2°C) at Vostok Station, Antarctica.

    "This record was set on July 21, 1983, demonstrating the extreme cold conditions that can occur at high latitudes."

- **Fact 33**: The Earth's atmosphere is composed of 78% nitrogen and 21% oxygen.

    "These two gases make up the majority of the atmosphere, with trace amounts of other gases like carbon dioxide and argon."

- **Fact 34**: The Coriolis effect causes winds to curve due to the Earth's rotation.

  "This phenomenon influences weather patterns and ocean currents, contributing to the formation of cyclones and trade winds."

- **Fact 35**: Hurricanes are the most powerful type of storm on Earth.

  "These intense tropical cyclones can cause widespread damage due to strong winds, heavy rain, and storm surges."

- **Fact 36**: Tornadoes are rapidly rotating columns of air extending from a thunderstorm to the ground.

  "Tornadoes can cause significant destruction with their high wind speeds and unpredictable paths."

- **Fact 37**: The ozone layer protects life on Earth from harmful ultraviolet (UV) radiation.

    "Located in the stratosphere, the ozone layer absorbs most of the Sun's harmful UV rays, preventing them from reaching the surface."

- **Fact 38**: El Niño and La Niña are climate phenomena that affect global weather patterns.

    "These events result from variations in ocean temperatures in the central and eastern Pacific, impacting precipitation, storms, and temperature."

- **Fact 39**: The Earth's tilt on its axis causes the seasons.

    "The 23.5-degree tilt results in varying sunlight intensity and day length throughout the year, creating seasonal changes."

- **Fact 40**: The water cycle helps regulate the Earth's climate.

    "By distributing heat and moisture, the water cycle plays a key role in maintaining the Earth's temperature balance."

- **Fact 41**: The Grand Canyon is over 270 miles long.

    "Carved by the Colorado River, this immense canyon showcases layers of geological history and attracts millions of visitors each year."

- **Fact 42**: Mount Everest is the highest peak above sea level.

  "Standing at 29,029 feet (8,848 meters), Everest is located in the Himalayas on the border between Nepal and China."

- **Fact 43**: The Great Barrier Reef is visible from space.

  "This massive coral reef system off the coast of Australia is one of the most biodiverse and extensive ecosystems on Earth."

- **Fact 44**: The Amazon Rainforest is the largest tropical rainforest in the world.

  "Covering approximately 5.5 million square kilometers, the Amazon Rainforest spans nine countries and is home to an incredible diversity of plant and animal species."

- **Fact 45**: The Sahara Desert is the largest hot desert in the world.

  "The Sahara, located in North Africa, covers an area of about 9.2 million square kilometers and is known for its vast stretches of sand dunes and extreme temperatures."

- **Fact 46**: Angel Falls is the world's highest uninterrupted waterfall.

  "Located in Venezuela, Angel Falls drops over 3,200 feet (979 meters) from the Auyán-Tepuí mountain, creating a stunning natural spectacle."

- **Fact 47**: The Dead Sea is the lowest point on the Earth's surface.

    "The shoreline of the Dead Sea is about 430 meters (1,411 feet) below sea level, making it the lowest land elevation on Earth."

- **Fact 48**: The Aurora Borealis, or Northern Lights, is a natural light display in the Earth's sky.

    "Caused by the interaction of solar wind with the Earth's magnetosphere, the Aurora Borealis is most commonly visible in high-latitude regions around the Arctic and Antarctic."

- **Fact 49**: The Galápagos Islands are known for their unique and diverse wildlife.

    "These islands, located off the coast of Ecuador, are famous for their endemic species and played a key role in Charles Darwin's theory of evolution."

- **Fact 50**: The Victoria Falls is one of the largest and most famous waterfalls in the world.

    "Located on the Zambezi River at the border of Zambia and Zimbabwe, Victoria Falls is known for its impressive width of 1,708 meters (5,604 feet) and height of 108 meters (354 feet)."

- **Fact 51**: The Great Rift Valley is a major geological and geographical feature in East Africa.

  "Extending over 6,000 kilometers from Lebanon to Mozambique, the Great Rift Valley is a significant tectonic feature with numerous lakes, volcanoes, and wildlife reserves."

- **Fact 52**: The Giant's Causeway is a natural formation of basalt columns in Northern Ireland.

  "This UNESCO World Heritage site was formed by volcanic activity around 50-60 million years ago, resulting in roughly 40,000 interlocking basalt columns."

- **Fact 53**: Yellowstone National Park sits on top of a supervolcano.

  "The Yellowstone Caldera, located in the western United States, is one of the largest active volcanic systems in the world, known for its geothermal features like geysers and hot springs."

- **Fact 54**: The Cliffs of Moher are one of Ireland's most visited natural attractions.

  "These dramatic cliffs, rising up to 214 meters (702 feet) above the Atlantic Ocean, stretch for about 14 kilometers along the western coast of Ireland."

- **Fact 55**: The Blue Hole is a giant marine sinkhole off the coast of Belize.

  "Measuring over 300 meters (984 feet) across and 124 meters (407 feet) deep, the Great Blue Hole is a popular site for divers and marine biologists."

- **Fact 56**: The Rocky Mountains stretch over 3,000 miles in North America.

  "This major mountain range extends from northernmost part of British Columbia in Canada to New Mexico in the United States, featuring diverse ecosystems and abundant wildlife."

- **Fact 57**: The Atacama Desert is the driest desert in the world.

  "Located in Chile, the Atacama Desert receives less than 1 millimeter of rain per year in some areas, with certain regions experiencing no recorded rainfall."

- **Fact 58**: The Danube River flows through 10 countries in Europe.

  "The Danube, Europe's second-longest river, passes through countries including Germany, Austria, Slovakia, Hungary, and Romania before emptying into the Black Sea."

- **Fact 59**: The Marble Caves in Chile are natural wonders formed by water erosion.

  "Located on General Carrera Lake, the Marble Caves feature stunning blue and white marble formations created by thousands of years of wave action."

- **Fact 60**: The Great Wall of China is the longest man-made structure in the world.

  "Stretching over 13,000 miles, the Great Wall was built over centuries to protect against invasions and raids and is an iconic symbol of China's historical engineering."

- **Fact 61**: The Ngorongoro Crater in Tanzania is the world's largest inactive volcanic caldera.

  "Formed about 2.5 million years ago, this UNESCO World Heritage site spans approximately 260 square kilometers and supports a diverse range of wildlife."

- **Fact 62**: The Fjords of Norway are deep, glacially carved valleys filled with seawater.

  "These stunning natural formations were created by glacial activity during the ice ages, resulting in steep, dramatic cliffs and deep blue waters."

- **Fact 63**: The Okavango Delta in Botswana is one of the world's largest inland deltas.

  "This unique delta is a haven for wildlife, forming a lush and vibrant oasis in the middle of the arid Kalahari Desert."

- **Fact 64**: The Giant Sequoia trees in California are the largest trees by volume.

  "Found in Sequoia National Park, these ancient trees can grow over 300 feet tall and have a trunk diameter of over 30 feet."

- **Fact 65**: The Sahara Desert can reach temperatures of up to 136°F (58°C).

  "Known for its scorching temperatures, the Sahara is one of the hottest places on Earth, with extreme daytime heat and cold nighttime temperatures."

- **Fact 66**: The Amazon Rainforest produces 20% of the world's oxygen.

  "Often referred to as the "lungs of the Earth," the Amazon's dense vegetation plays a crucial role in the global oxygen cycle."

- **Fact 67**: The Waitomo Caves in New Zealand are famous for their glowworms.

  "These caves are illuminated by thousands of bioluminescent glowworms, creating a magical and unique underground light display."

- **Fact 68**: The Maldives is the lowest country in the world.

  "The average ground level is just 1.5 meters (4 feet 11 inches) above sea level, making the Maldives particularly vulnerable to rising sea levels."

- **Fact 69**: The Grand Prismatic Spring in Yellowstone is the largest hot spring in the United States.

  "This vibrant hot spring is known for its striking colors, caused by microbial mats around the edges of the mineral-rich water."

- **Fact 70**: The Great Basin Bristlecone Pine trees are the oldest living non-clonal organisms.

    "These ancient trees, found in the White Mountains of California, can live for over 5,000 years, making them some of the oldest living things on Earth."

- **Fact 71**: The Bay of Fundy has the highest tides in the world.

    "Located between New Brunswick and Nova Scotia in Canada, the Bay of Fundy experiences tidal changes of up to 16 meters (53 feet)."

- **Fact 72**: The Salar de Uyuni in Bolivia is the world's largest salt flat.

    "Covering over 10,000 square kilometers, this vast, reflective salt flat creates stunning mirror-like landscapes after rainfall."

- **Fact 73**: The Cave of Crystals in Mexico contains some of the largest natural crystals ever found.

    "Discovered in 2000, the cave features massive selenite crystals, some measuring over 11 meters (36 feet) long."

- **Fact 74**: The Iguazu Falls on the border of Argentina and Brazil are among the largest waterfall systems in the world.

    "Consisting of around 275 individual waterfalls, Iguazu Falls spans nearly 2.7 kilometers (1.7 miles) and is a UNESCO World Heritage site."

- **Fact 75**: The Great Barrier Reef supports over 1,500 species of fish.

    "This diverse ecosystem is home to thousands of marine species, including fish, mollusks, and marine mammals, making it a biodiversity hotspot."

- **Fact 76**: The Black Forest in Germany is famous for its dense woods and folklore.

    "This forested region inspired many traditional fairy tales and is known for its picturesque landscapes and cultural heritage."

- **Fact 77**: The Namib Desert in Namibia is one of the oldest deserts in the world.

    "Estimated to be around 55 million years old, the Namib Desert features towering sand dunes and unique desert-adapted wildlife."

- **Fact 78**: The Cliffs of Bandiagara in Mali are a UNESCO World Heritage site.

    "These dramatic cliffs and plateaus are home to the Dogon people and feature ancient dwellings and cultural sites."

- **Fact 79**: The Dolomites in Italy are known for their unique limestone formations.

    "This mountain range is characterized by its rugged peaks and dramatic cliffs, attracting climbers and nature enthusiasts from around the world."

- **Fact 80**: The Serengeti in Tanzania is famous for its annual wildebeest migration.

    "Each year, over 1.5 million wildebeest, along with hundreds of thousands of other animals, migrate in search of fresh grazing grounds."

- **Fact 81**: The Mendenhall Ice Caves in Alaska are located inside a glacier.

    "These stunning blue ice caves are formed within the Mendenhall Glacier, creating a surreal and otherworldly environment."

- **Fact 82**: The Matterhorn is one of the highest peaks in the Alps.

    "Located on the border between Switzerland and Italy, the Matterhorn stands at 4,478 meters (14,692 feet) and is one of the most iconic mountains in the world."

- **Fact 83**: The Carlsbad Caverns in New Mexico contain over 119 caves.

  "Formed by sulfuric acid dissolution, these extensive caverns feature impressive limestone formations and underground chambers."

- **Fact 84**: The Drakensberg Mountains in South Africa are a UNESCO World Heritage site.

  "Known for their dramatic cliffs and stunning landscapes, the Drakensberg Mountains are home to diverse flora and fauna."

- **Fact 85**: The Plitvice Lakes in Croatia are a series of interconnected lakes and waterfalls.

  "This national park features 16 terraced lakes, known for their distinctive colors and crystal-clear waters, connected by cascades and waterfalls."

- **Fact 86**: The Uluru-Kata Tjuta National Park in Australia is home to the iconic Uluru rock formation.

  "Also known as Ayers Rock, Uluru is a massive sandstone monolith sacred to the Aboriginal people and a symbol of Australia's natural heritage."

- **Fact 87**: The Wulingyuan Scenic Area in China is known for its sandstone pillars.

  "This UNESCO World Heritage site features thousands of towering sandstone pillars and is famous for its breathtaking scenery and biodiversity."

- **Fact 88**: The Chocolate Hills in the Philippines are a geological formation of over 1,200 mounds.

    "These unique hills turn brown during the dry season, resembling chocolate drops, and are a major tourist attraction in the Bohol province."

- **Fact 89**: The Pamukkale terraces in Turkey are made of travertine.

    "These stunning white terraces are formed by mineral-rich thermal waters, creating a surreal landscape of cascading pools."

- **Fact 90**: The Waitomo Glowworm Caves in New Zealand are illuminated by bioluminescent larvae.

    "These caves are famous for their glowworm displays, where thousands of bioluminescent larvae create a starry-night effect on the cave ceilings."

- **Fact 91**: The Moeraki Boulders in New Zealand are unusually large spherical stones.

    "These spherical boulders, found on Koekohe Beach, formed millions of years ago through calcite deposition and erosion."

- **Fact 92**: The Marble Cathedral in Chile is a stunning series of caves carved by water.

    "Located on General Carrera Lake, these marble caves have been shaped by wave action over thousands of years, creating intricate patterns and colors."

- **Fact 93**: The Petrified Forest in Arizona is known for its fossilized trees.

    "This national park contains large deposits of petrified wood, formed over 200 million years ago, offering a glimpse into ancient ecosystems."

- **Fact 94**: The Jeita Grotto in Lebanon is a system of interconnected limestone caves.

    "These caves feature stunning stalactites and stalagmites, as well as an underground river that can be explored by boat."

- **Fact 95**: The White Desert in Egypt is famous for its chalk rock formations.

    "This desert is known for its surreal white chalk formations, sculpted by wind and sand into unique shapes and figures."

- **Fact 96**: The Valley of the Ten Peaks in Canada offers stunning mountain scenery.

    "Located in Banff National Park, this valley is surrounded by ten towering peaks and is known for its breathtaking landscapes and turquoise lakes."

- **Fact 97**: The Fjallabak Nature Reserve in Iceland features colorful rhyolite mountains.

  "This reserve is known for its strikingly colorful mountains, geothermal activity, and scenic hiking trails."

- **Fact 98**: The Iguazu National Park is home to one of the world's largest waterfall systems.

  "Located on the border of Argentina and Brazil, Iguazu National Park features around 275 individual waterfalls and is a UNESCO World Heritage site."

- **Fact 99**: The Mount Roraima in Venezuela is one of the world's highest table mountains.

  "This flat-topped mountain, known as a tepui, stands at 2,810 meters (9,219 feet) and is surrounded by sheer cliffs and unique ecosystems."

- **Fact 100**: The Ngorongoro Crater in Tanzania is a UNESCO World Heritage site.

  "Formed by a collapsed volcano, the Ngorongoro Crater is one of the best places in Africa to see wildlife, including the "Big Five" game animals."

- **Fact 101**: The Earth's magnetic field is generated by its outer core.

    "The movement of molten iron and nickel in the Earth's outer core creates electric currents, which in turn generate the planet's magnetic field."

- **Fact 102**: The Amazon Rainforest is often referred to as the "lungs of the Earth.

    "This vast rainforest produces about 20% of the world's oxygen through photosynthesis."

- **Fact 103**: Mount Kilimanjaro is the highest peak in Africa.

    "Located in Tanzania, Mount Kilimanjaro stands at 19,341 feet (5,895 meters) above sea level."

- **Fact 104**: The deepest lake in the world is Lake Baikal in Russia.

    "Lake Baikal reaches a maximum depth of about 5,387 feet (1,642 meters) and holds 20% of the world's unfrozen freshwater."

- **Fact 105**: The world's largest desert is Antarctica.

    "Despite being covered in ice, Antarctica is classified as a desert due to its extremely low precipitation levels.

- **Fact 106**: The Earth's atmosphere extends up to about 10,000 kilometers (6,200 miles) above the planet.

  "This layer of gases protects life on Earth by absorbing ultraviolet solar radiation and reducing temperature extremes."

- **Fact 107**: The Mariana Trench is the deepest part of the world's oceans.

  "Located in the western Pacific Ocean, the Mariana Trench reaches a depth of about 36,070 feet (10,994 meters) at its deepest point, known as Challenger Deep."

- **Fact 108**: The Earth rotates on its axis at approximately 1,000 miles per hour (1,600 kilometers per hour) at the equator.

  "This rotation is responsible for the cycle of day and night."

- **Fact 109**: The Earth's largest ice sheet is located in Antarctica.

  "The Antarctic Ice Sheet covers about 14 million square kilometers and contains around 60% of the world's fresh water."

- **Fact 110**: The phenomenon of bioluminescence occurs in some marine organisms.

  "Certain marine species, like jellyfish and plankton, produce light through chemical reactions within their bodies, creating a glowing effect.

- **Fact 111**: The Earth's surface is composed of about 71% water and 29% land.

   "This distribution of water and land is essential for sustaining life and regulating the planet's climate."

- **Fact 112**: The Dead Sea is nearly 10 times saltier than the ocean.

   "The high salinity of the Dead Sea makes it one of the saltiest bodies of water on Earth, preventing most marine life from surviving there."

- **Fact 113**: The Earth's atmosphere contains trace amounts of carbon dioxide, which is crucial for photosynthesis.

   "Carbon dioxide is a vital component of the carbon cycle, allowing plants to produce oxygen through photosynthesis."

- **Fact 114**: The Earth's longest mountain range is the Mid-Atlantic Ridge.

   "This underwater mountain range extends for about 16,000 kilometers (10,000 miles) along the floor of the Atlantic Ocean."

- **Fact 115**: The Earth experiences about 100,000 thunderstorms each year.

   "Thunderstorms are common weather events that can occur anywhere in the world, with lightning, heavy rain, and strong winds."

- **Fact 116**: ThThe Himalayas are the youngest mountain range on Earth.

    "Formed by the collision of the Indian and Eurasian tectonic plates, the Himalayas are still rising and are geologically young compared to other mountain ranges."

- **Fact 117**: The Earth's biosphere includes all living organisms and their physical environments.

    "The biosphere encompasses ecosystems on land, in the oceans, and in the atmosphere, supporting a diverse range of life forms."

- **Fact 118**: The Earth's core is composed mainly of iron and nickel.

    "These metals are responsible for the core's high density and are crucial for generating the Earth's magnetic field."

- **Fact 119**: The Great Victoria Desert is the largest desert in Australia.

    "Covering an area of about 348,750 square kilometers, this desert features sand dunes, salt lakes, and a variety of desert flora and fauna."

- **Fact 120**: The deepest cave in the world is the Veryovkina Cave in Georgia.

    "The Veryovkina Cave, located in the Caucasus Mountains, has a depth of about 7,257 feet (2,212 meters), making it the deepest known cave on Earth."

- **Fact 121**: The Earth's largest living structure is the Great Barrier Reef.

    "The Great Barrier Reef, located off the coast of Australia, is the largest living structure on Earth, stretching over 1,400 miles (2,300 kilometers) and composed of billions of tiny coral polyps."

- **Fact 122**: The longest river in the world is the Nile River.

    "The Nile River, stretching about 4,135 miles (6,650 kilometers) through northeastern Africa, is traditionally considered the longest river in the world, flowing through 11 countries and providing vital resources to millions of people."

Earth Trivia Questions-1
(Answers at the end)

Question 1: What is the thinnest layer of the Earth?
A) Crust
B) Mantle
C) Outer Core
D) Inner Core

Question 2: Which of the following is the largest ocean on Earth?
A) Atlantic Ocean
B) Indian Ocean
C) Arctic Ocean
D) Pacific Ocean

Question 3: What is the primary gas in the Earth's atmosphere?
A) Oxygen
B) Nitrogen
C) Carbon Dioxide
D) Argon

Question 4: Which natural phenomenon is measured using the Richter scale?
A) Hurricanes
B) Tornadoes
C) Earthquakes
D) Volcanic eruptions

Question 5: The Ring of Fire is located in which ocean?
A) Atlantic Ocean
B) Indian Ocean
C) Arctic Ocean
D) Pacific Ocean

Question 6: What is the deepest point in the world's oceans?
A) Tonga Trench
B) Mariana Trench
C) Java Trench
D) Puerto Rico Trench

Question 7: Which of the following deserts is the largest hot desert in the world?
A) Gobi Desert
B) Sahara Desert
C) Kalahari Desert
D) Atacama Desert

Question 8: The Great Barrier Reef is located off the coast of which country?
A) United States
B) Brazil
C) Australia
D) South Africa

Earth Trivia Questions-2
(Answers at the end)

Question 9: Which river is the largest by discharge volume?

A) Nile River
B) Mississippi River
C) Amazon River
D) Yangtze River

Question 10: Which is the highest peak above sea level?

A) K2
B) Mount Kilimanjaro
C) Mount Everest
D) Denali

Question 11: The lowest point on the Earth's surface is located at which body of water?

A) Caspian Sea
B) Dead Sea
C) Great Salt Lake
D) Lake Baikal

Question 12: Which of the following is the largest coral reef system in the world?

A) Andros Barrier Reef
B) Belize Barrier Reef
C) Red Sea Coral Reef
D) Great Barrier Reef

Question 13: What is the main cause of tsunamis?

A) Hurricanes
B) Tornadoes
C) Underwater earthquakes
D) Volcanic eruptions

Question 14: Which ocean current influences the climate of the east coast of North America and Europe?

A) Kuroshio Current
B) California Current
C) Gulf Stream
D) Antarctic Circumpolar Current

Earth Trivia Questions-3
(Answers at the end)

Question 15: What natural feature is Angel Falls known for?

A) It is the highest waterfall in the world.
B) It is the widest waterfall in the world.
C) It is the most powerful waterfall in the world.
D) It is the longest waterfall in the world.

Question 16: Which continent is home to the Amazon Rainforest?

A) Africa
B) Asia
C) South America
D) Australia

Question 17: What is the primary component of Saturn's rings?

A) Rock
B) Ice
C) Dust
D) Metal

Question 18: The Cliffs of Moher are located in which country?

A) Scotland
B) Norway
C) Ireland
D) Iceland

Question 19: The phenomenon of the Northern Lights is also known as what?

A) Solar Wind
B) Aurora Borealis
C) Equinox
D) Midnight Sun

Question 20: Which desert is known as the driest place on Earth?

A) Gobi Desert
B) Sahara Desert
C) Atacama Desert
D) Mojave Desert

Introduction to Animals

Welcome to the captivating world of animals! Our planet is teeming with an incredible diversity of animal life, each species uniquely adapted to its environment. In this section, we will explore fascinating facts about mammals, birds, reptiles, amphibians, fish, and insects. Get ready to be amazed by the wonders of the animal kingdom!

- **Fact 1**: The blue whale is the largest animal on Earth..

    "Blue whales can reach lengths of up to 100 feet (30 meters) and weigh as much as 200 tons. Their heart alone can be as large as a small car."

- **Fact 2**: Bats are the only mammals capable of sustained flight.

    "While other mammals like flying squirrels can glide, bats are unique in their ability to fly continuously, thanks to their specially adapted wings."

- **Fact 3**: The platypus is one of the few mammals that lay eggs.

    "Found in Australia, the platypus is a monotreme, a primitive group of egg-laying mammals. The echidna is another example."

- **Fact 4**: Kangaroos can jump up to 30 feet in a single leap.

    "These marsupials use their powerful hind legs and long tails for balance, allowing them to travel great distances quickly."

- **Fact 5**: Dolphins are highly intelligent marine mammals.

    "Dolphins have large brains relative to their body size and exhibit complex behaviors, including problem-solving, communication, and social interaction."

- **Fact 6**: Elephants have the longest gestation period of any land animal.

    "Female elephants are pregnant for about 22 months, giving birth to calves that can weigh around 250 pounds at birth."

- **Fact 7**: A group of lions is called a pride.

    "Lions are the only big cats that live in social groups, with prides typically consisting of related females, their offspring, and a few males."

- **Fact 8**: Polar bears have black skin under their white fur.

    "The black skin absorbs heat from the sun, helping polar bears stay warm in their Arctic habitat. Their white fur provides camouflage against the snow and ice."

- **Fact 9**: The cheetah is the fastest land animal, capable of reaching speeds up to 70 mph.

    "Cheetahs use their incredible speed to chase down prey, with acceleration unmatched by any other land animal."

- **Fact 10**: Koalas sleep up to 18-22 hours a day.

    "Koalas have a low-energy diet of eucalyptus leaves, which require a lot of energy to digest, leading to their long sleep periods."

- **Fact 11**: The ostrich is the largest bird in the world.

    "Ostriches can grow up to 9 feet tall and weigh as much as 320 pounds. They are also the fastest running birds, capable of sprinting at speeds up to 45 mph."

- **Fact 12**: Hummingbirds can fly backwards.

    "These tiny birds are the only ones capable of sustained backward flight, thanks to their unique wing structure and rapid flapping."

- **Fact 13**: Penguins are flightless birds that excel at swimming.

    "Penguins' wings have evolved into flippers, allowing them to navigate the icy waters of the Southern Hemisphere with agility and speed."

- **Fact 14**: The harpy eagle has the strongest talons of any bird of prey.

    "Native to Central and South American rainforests, the harpy eagle's powerful talons can exert pressure of up to 530 psi, enabling it to catch large prey."

- **Fact 15**: The albatross has the longest wingspan of any bird.

    "With wingspans reaching up to 12 feet, albatrosses can glide effortlessly over the ocean for hours without flapping their wings."

- **Fact 16**: The kiwi is a nocturnal bird with nostrils at the end of its beak.

  "Found in New Zealand, kiwis use their highly developed sense of smell to locate insects and worms at night."

- **Fact 17**: Peacocks are known for their vibrant tail feathers, which they fan out in a display.

  "Male peacocks use their colorful tail feathers to attract females during mating season, creating a stunning visual display."

- **Fact 18**: The peregrine falcon is the fastest bird, capable of diving at speeds over 200 mph.

  "During a hunting dive, called a stoop, peregrine falcons can reach incredible speeds, making them formidable predators."

- **Fact 19**: The Arctic tern migrates farther than any other bird.

  "Arctic terns travel from their breeding grounds in the Arctic to the Antarctic and back each year, covering a round-trip distance of about 44,000 miles."

- **Fact 20**: The lyrebird is known for its extraordinary ability to mimic sounds.

  "Native to Australia, lyrebirds can imitate natural and artificial sounds, including chainsaws, camera shutters, and other birds' calls."

- **Fact 21**: The leatherback sea turtle is the largest turtle in the world.

  "Leatherbacks can grow up to 7 feet long and weigh over 2,000 pounds. They are unique among sea turtles for their lack of a hard shell."

- **Fact 22**: Crocodiles can live up to 70-100 years.

  "These ancient reptiles have remarkable longevity, with some individuals living for over a century in the wild."

- **Fact 23**: Chameleons can change color to communicate and regulate their temperature.

  "Contrary to popular belief, chameleons primarily change color for social signaling and temperature control rather than camouflage."

- **Fact 24**: The green anaconda is the heaviest snake in the world.

    "Found in South America, green anacondas can weigh over 500 pounds and reach lengths of up to 30 feet."

- **Fact 25**: The Komodo dragon is the largest living lizard.

    "Native to Indonesia, Komodo dragons can grow up to 10 feet long and weigh around 150 pounds. They are powerful predators with venomous bites."

- **Fact 26**: Geckos can climb smooth surfaces due to tiny hairs on their feet.

    "These microscopic hairs, called setae, create van der Waals forces that allow geckos to adhere to walls and ceilings."

- **Fact 27**: Some species of snakes give birth to live young instead of laying eggs.

  "Known as viviparous snakes, these species, including boas and vipers, give birth to fully formed, live offspring."

- **Fact 28**: Tortoises are among the longest-lived animals on Earth.

  "Certain species, such as the Galápagos tortoise, can live for over 150 years, making them some of the longest-living vertebrates."

- **Fact 29**: The Gila monster is one of the few venomous lizards.

  "Found in the southwestern United States and Mexico, the Gila monster has venom glands in its lower jaw that it uses to subdue prey."

- **Fact 30**: The tuatara, a reptile native to New Zealand, has a third "parietal" eye.

  "This primitive eye, located on the top of the head, is thought to be involved in regulating circadian rhythms and detecting light."

- **Fact 31**: The axolotl can regenerate lost body parts.

  "This unique salamander, native to Mexico, has the remarkable ability to regrow limbs, spinal cord, heart, and other organs."

- **Fact 32**: Frogs can absorb water through their skin.

  "Frogs have permeable skin that allows them to absorb moisture directly from their environment, eliminating the need to drink water."

- **Fact 33**: The poison dart frog's skin contains potent toxins.

  "These brightly colored frogs produce toxins that can be lethal to predators. Indigenous people have used their poison to coat blow darts for hunting."

- **Fact 34**: Salamanders can breathe through their skin.

  "Many salamanders have thin, moist skin that allows them to absorb oxygen and release carbon dioxide directly from their environment."

- **Fact 35**: The largest amphibian is the Chinese giant salamander.

  "This critically endangered species can grow up to 6 feet long and weigh over 140 pounds."

- **Fact 36**: The Surinam toad gives birth through its back.

  "Female Surinam toads have specialized skin pockets on their backs where they incubate and hatch their eggs."

- **Fact 37**: The hellbender is the largest salamander in North America.

    "This aquatic salamander can grow up to 29 inches long and is known for its flattened body and wrinkled skin."

- **Fact 38**: Newts can regenerate their spinal cords.

    "Newts possess remarkable regenerative abilities, including the ability to regrow spinal cord tissue, limbs, and even parts of their heart and eyes."

- **Fact 39**: The glass frog has translucent skin.

    "These frogs have skin so translucent that their internal organs, including the heart and intestines, are visible through their belly."

- **Fact 40**: The mudpuppy is a fully aquatic salamander that retains its gills throughout life.

    "Unlike many amphibians that lose their gills during development, the mudpuppy keeps its external gills and lives entirely in water."

- **Fact 41**: The whale shark is the largest fish in the world.

    "Whale sharks can grow up to 40 feet long and weigh as much as 20.6 tons. Despite their size, they are filter feeders and primarily eat plankton."

- **Fact 42**: Some fish can change their sex.

    "Species like clownfish and wrasses can change their sex in response to environmental or social cues, often as a survival strategy."

- **Fact 43**: The electric eel can produce an electric shock of up to 600 volts.

    "This South American fish uses its electric organs to stun prey and defend against predators."

- **Fact 44**: The coelacanth was thought to be extinct until it was rediscovered in 1938.

    "This ancient fish, often called a "living fossil," was believed to have gone extinct 66 million years ago until a live specimen was found off the coast of South Africa."

- **Fact 45**: The pufferfish can inflate itself to avoid predators.

  "When threatened, pufferfish ingest large amounts of water (or air) to balloon up, making themselves harder for predators to swallow."

- **Fact 46**: The anglerfish uses a bioluminescent lure to attract prey.

  "Deep-sea anglerfish have a glowing lure on their heads, produced by symbiotic bacteria, to entice prey in the dark depths of the ocean."

- **Fact 47**: Some species of fish communicate with sound.

  "Fish like the croaker and the drumfish produce sounds using specialized muscles and swim bladders to communicate with each other."

- **Fact 48**: The parrotfish can change color throughout its life.

  "Parrotfish undergo color changes as they grow, which can signal different stages of maturity or social status."

- **Fact 49**: The goblin shark has a distinctive, elongated snout.

  "This deep-sea shark uses its long, flat snout to sense prey in the dark ocean depths and can extend its jaws forward to catch it."

- **Fact 50**: The seahorse is the only fish species in which males give birth.

    "Male seahorses have a brood pouch where females deposit their eggs. The males then fertilize and carry the eggs until they hatch."

- **Fact 51**: There are more species of beetles than any other type of animal.

    "Beetles make up about 25% of all known animal species, with over 400,000 identified species worldwide."

- **Fact 52**: Honeybees communicate through a "waggle dance."

    "Honeybees perform a waggle dance to share information about the direction and distance to food sources with other members of their hive."

- **Fact 53**: Dragonflies can fly in all directions, including backward.

  "Dragonflies have two sets of wings that operate independently, allowing them to hover, fly backward, and change direction rapidly."

- **Fact 54**: Ants can lift objects up to 50 times their body weight.

  "Ants are incredibly strong for their size, using their powerful mandibles to carry heavy loads back to their colonies."

- **Fact 55**: The monarch butterfly migrates up to 3,000 miles each year.

  "Monarch butterflies travel from North America to central Mexico for the winter, making one of the longest migrations of any insect."

- **Fact 56**: Fireflies produce light through a chemical reaction called bioluminescence.

  "This natural light production is used for communication, mating, and warding off predators."

- **Fact 57**: The praying mantis has excellent eyesight and can turn its head 180 degrees.

  "Praying mantises have two large compound eyes and three simple eyes, giving them a wide field of vision to spot prey and predators."

- **Fact 58**: Termites are important decomposers in many ecosystems.

  "Termites break down cellulose in dead wood and plant material, recycling nutrients back into the soil and supporting plant growth."

- **Fact 59**: The atlas moth has the largest wingspan of any moth, reaching up to 12 inches.

  "Found in Southeast Asia, the atlas moth's impressive wingspan makes it one of the largest insects in the world."

- **Fact 60:** Mosquitoes are attracted to humans by the carbon dioxide we exhale.

  "In addition to carbon dioxide, mosquitoes are drawn to body heat and certain chemicals found in sweat."

- **Fact 61**: Some frogs can freeze without dying.

  "Certain frog species can survive freezing temperatures by entering a state of suspended animation, where their bodily functions temporarily shut down until they thaw."

- **Fact 62**: The mudskipper can breathe through its skin and gills, allowing it to live on land and in water.

  "Mudskippers are amphibious fish that use their pectoral fins to move on land and breathe through their skin when out of water."

- **Fact 63**: The lungfish can survive out of water for several years.

  "Lungfish have both gills and lungs, allowing them to breathe air and survive droughts by burying themselves in mud and entering a state of hibernation."

- **Fact 64**: The clownfish has a mutualistic relationship with sea anemones.

  "Clownfish live among the venomous tentacles of sea anemones, gaining protection from predators while providing the anemone with food scraps."

- **Fact 65**: The archerfish can shoot jets of water to knock insects off leaves.

  "Archerfish have a specialized mouth structure that allows them to accurately shoot water at prey, bringing it down into the water to be eaten."

- **Fact 66**: The blobfish has a gelatinous body that allows it to withstand high pressure.

    "Found in deep ocean waters, the blobfish's jelly-like body provides buoyancy and helps it survive the extreme pressure of its environment."

- **Fact 67**: The stonefish is one of the most venomous fish in the world.

    "Stonefish have venomous spines that can deliver a painful and potentially lethal sting to predators and humans."

- **Fact 68**: The flying fish can glide above the water's surface to escape predators.

    "These fish use their long, wing-like pectoral fins to leap out of the water and glide for short distances, evading aquatic predators."

- **Fact 69**: The clown loach can make clicking sounds when it is happy.

    "This behavior, often observed during feeding or social interactions, is thought to be a form of communication among clown loaches."

- **Fact 70**: The betta fish, or Siamese fighting fish, is known for its aggressive behavior and vibrant colors.

    "Male bettas are highly territorial and will fight other males, displaying their brightly colored fins to assert dominance."

- **Fact 71**: The Hercules beetle is one of the largest beetles in the world.

    "Male Hercules beetles can reach lengths of up to 7 inches, including their long, curved horns used for fighting rivals."

- **Fact 71**: The bombardier beetle can eject a boiling, noxious chemical spray to deter predators.

    "This beetle has specialized glands that mix chemicals to produce a hot, irritating spray as a defense mechanism."

- **Fact 73**: The leafcutter ant is capable of carrying pieces of leaves many times its own body weight.

    "Leafcutter ants use their powerful jaws to cut and transport leaf fragments back to their colonies, where the leaves are used to cultivate fungus for food."

- **Fact 74**: The assassin bug injects enzymes to liquefy its prey's insides.

    "These predatory insects use their long, sharp mouthparts to inject digestive enzymes into their prey, then suck out the liquefied contents."

- **Fact 75**: The praying mantis has a specialized hunting behavior called "cryptic mimicry."

    "Praying mantises can blend in with their surroundings by mimicking the appearance of leaves, flowers, or twigs, allowing them to ambush unsuspecting prey."

- **Fact 76**: The dung beetle is known for its role in recycling animal waste.

    "Dung beetles collect and bury animal feces, which they use as a food source and a place to lay their eggs, helping to decompose organic matter and enrich the soil."

- **Fact 77**: The atlas moth does not have a functional mouth and does not eat as an adult.

    "After emerging from its cocoon, the atlas moth relies on stored energy from its larval stage and lives for only a few days to a week."

- **Fact 78**: The cicada has one of the longest lifecycles of any insect, with some species emerging every 13 or 17 years.

    "Periodical cicadas spend most of their lives underground as nymphs, emerging in massive numbers to mate and lay eggs."

- **Fact 79**: The tarantula hawk wasp has one of the most painful stings in the insect world.

    "This large wasp preys on tarantulas, using its powerful sting to paralyze the spider before laying eggs on it as a food source for its larvae."

- **Fact 80:** The leaf insect is a master of camouflage, resembling a leaf in both shape and color.

    "This incredible mimicry helps leaf insects avoid predators by blending in perfectly with their natural environment."

- **Fact 81**: The wombat has cube-shaped poop.

    "This unusual shape prevents the poop from rolling away, helping wombats mark their territory and communicate with each other."

- **Fact 82**: The horned lizard can squirt blood from its eyes as a defense mechanism..

    "When threatened, the horned lizard can rupture blood vessels in its eyes, spraying blood at predators to deter them."

- **Fact 83**: The narwhal's tusk is actually a long tooth.

    "Male narwhals have a single elongated tooth that can grow up to 10 feet long, which is used for sensing the environment and possibly for mating displays."

- **Fact 84**: The aye-aye is a nocturnal lemur with a long middle finger used for hunting insects.

  "This unique primate taps on tree bark to find hollow spots where insects hide, then uses its elongated finger to extract them."

- **Fact 85**: The bowerbird builds elaborate structures, called bowers, to attract mates.

  "Male bowerbirds collect colorful objects and arrange them around their bowers to impress females and increase their chances of mating."

- **Fact 86**: The hoatzin, a bird native to South America, has a unique digestive system similar to that of a cow.

  "The hoatzin's enlarged crop ferments leaves and other vegetation, allowing it to extract nutrients in a process similar to ruminants."

- **Fact 87**: The Gharial, a crocodile-like reptile, has a long, narrow snout adapted for catching fish.

  "Found in the rivers of the Indian subcontinent, the Gharial's specialized snout and sharp teeth make it an efficient fish hunter."

- **Fact 88**: The Capuchin monkey uses tools to crack open nuts.

  "These intelligent primates are known for their ability to use rocks and other tools to access food, demonstrating advanced problem-solving skills."

- **Fact 89**: The Wandering Albatross has the longest wingspan of any living bird.

  "With wingspans reaching up to 12 feet, the Wandering Albatross can glide for hours over the ocean without flapping its wings."

- **Fact 90**: The mantis shrimp has one of the fastest punches in the animal kingdom.

  "Thages."

- **Fact 91**: The octopus has three hearts and blue blood.

  "Two hearts pump blood to the gills, while the third pumps it to the rest of the body. Their blue blood contains a copper-based molecule called hemocyanin, which is more efficient at transporting oxygen in cold, low-oxygen environments."

- **Fact 92**: The mimic octopus can imitate other sea creatures.

    "This clever octopus can change its color, shape, and behavior to mimic various marine animals like lionfish, flatfish, and sea snakes to avoid predators."

- **Fact 93**: Flamingos are born grey and turn pink from their diet.

    "Flamingos' pink color comes from carotenoid pigments in the algae and crustaceans they eat, which is broken down into pink pigments by their bodies."

- **Fact 94**: Male seahorses give birth to their young.

    "Male seahorses have a brood pouch where females deposit their eggs. The males then fertilize, carry, and give birth to the baby seahorses."

- **Fact 95**: Some species of jellyfish are considered immortal.

    "The Turritopsis dohrnii jellyfish can revert to its juvenile form after reaching adulthood, essentially bypassing death and starting its life cycle anew."

- **Fact 96**: The peregrine falcon is the fastest bird, capable of diving at speeds over 200 mph.

    "During a hunting dive, called a stoop, peregrine falcons can reach incredible speeds, making them formidable predators."

- **Fact 97**: The axolotl can regenerate lost body parts.

    "This unique salamander, native to Mexico, has the remarkable ability to regrow limbs, spinal cord, heart, and other organs."

- **Fact 98**: The blue dragon sea slug floats on the ocean's surface and preys on venomous creatures.

    "This tiny but stunning sea slug can consume venomous prey like the Portuguese man o' war and store their toxins to use for its own defense."

- **Fact 99**: Starfish can regrow lost arms.

    "If a starfish loses an arm, it can regenerate it. Some species can even regenerate a whole new starfish from a single lost arm if it contains part of the central disc."

- **Fact 100**: The lyrebird is known for its extraordinary ability to mimic sounds.

    "Native to Australia, lyrebirds can imitate natural and artificial sounds, including chainsaws, camera shutters, and other birds' calls."

- **Fact 101**: The mimic poison dart frog changes its appearance to resemble other toxic species.

    "This frog uses mimicry to avoid predators, imitating the appearance of more toxic species to deter potential threats."

- **Fact 102**: The ghost shrimp is nearly transparent.

  "Ghost shrimp have see-through bodies that help them avoid predators by blending into their surroundings in the water."

- **Fact 103**: The dragonfly nymph can shoot water out of its rectum to propel itself.

  "Dragonfly nymphs use this jet propulsion method to escape predators quickly while living in water during their larval stage."

- **Fact 104**: The secretary bird hunts snakes by stomping on them.

  "This large bird of prey, found in Africa, uses its powerful legs to stomp on snakes and other prey, delivering lethal blows."

- **Fact 105**: The sea cucumber can eject its internal organs as a defense mechanism.

  "When threatened, sea cucumbers can expel their guts to entangle and distract predators, later regenerating the lost organs."

- **Fact 106**: The pink fairy armadillo is the smallest species of armadillo.

  "Native to central Argentina, this tiny armadillo grows to about 6 inches long and has a pinkish shell that helps it blend into its sandy environment."

- **Fact 107**: The saiga antelope has a distinctive, bulbous nose.

    "This critically endangered antelope from the Eurasian steppes has a unique nose that helps filter dust and regulate its blood temperature."

- **Fact 108**: The red-lipped batfish has a unique appearance with bright red lips.

    "Found around the Galápagos Islands, this unusual fish uses its pectoral fins to "walk" on the ocean floor and is known for its distinctive red lips."

- **Fact 109**: The electric catfish can produce a shock of up to 450 volts.

    "Native to Africa, the electric catfish uses its electrical abilities to stun prey and defend against predators, similar to the electric eel."

- **Fact 110**: .

    "The heart of a blue whale, the largest animal on Earth, can weigh approximately 400 pounds and is about the size of a small car. This massive heart pumps blood through the whale's enormous body, supporting its vital functions and allowing it to thrive in the ocean."

Animal Trivia Questions-1
(Answers at the end)

Question 1: What is the largest animal on Earth?
A) Elephant
B) Great White Shark
C) Blue Whale
D) Giraffe

Question 2: Which mammal is capable of sustained flight?
A) Bat
B) Flying Squirrel
C) Flying Lemur
D) Sugar Glider

Question 3: What unique feature do male seahorses have?
A) They have a pouch for carrying eggs
B) They can change color
C) They can regenerate limbs
D) They produce ink

Question 4: Which bird is known for its ability to mimic sounds?
A) Lyrebird
B) Parrot
C) Crow
D) Nightingale

Question 5: Which reptile has a third "parietal" eye?
A) Komodo Dragon
B) Tuatara
C) Gecko
D) Chameleon

Question 6: What is the fastest land animal?
A) Cheetah
B) Lion
C) Greyhound
D) Horse

Question 7: How do flamingos get their pink color?
A) From the algae and crustaceans they eat
B) From exposure to the sun
C) From their natural genetic makeup
D) From mud baths

Question 8: Which fish can produce an electric shock of up to 600 volts?
A) Electric Catfish
B) Electric Eel
C) Stingray
D) Piranha

Animal Trivia Questions-2
(Answers at the end)

Question 9: Which insect communicates through a "waggle dance"?

A) Ant
B) Butterfly
C) Honeybee
D) Beetle

Question 10: What is the largest species of turtle?

A) Green Sea Turtle
B) Leatherback Sea Turtle
C) Loggerhead Turtle
D) Hawksbill Turtle

Question 11: Which amphibian can regenerate its limbs and other body parts?

A) Frog
B) Salamander
C) Newt
D) Axolotl

Question 12: Which bird has the longest wingspan?

A) Eagle
B) Condor
C) Albatross
D) Pelican

Question 13: What defense mechanism does the sea cucumber use?

A) Ejects ink
B) Ejects internal organs
C) Changes color
D) Inflates body

Question 14: How do mudskippers breathe when on land?

A) Through their gills
B) Through their skin
C) Through their lungs
D) Through their fins

Animal Trivia Questions-3
(Answers at the end)

Question 15: Which bird can fly backwards?

A) Sparrow
B) Eagle
C) Hummingbird
D) Pigeon

Question 16: What is unique about the reproductive process of clownfish?

A) Males give birth
B) Females change color
C) Clownfish can change sex
D) They lay eggs on coral

Question 17: What is the main diet of a blue whale?

A) Fish
B) Plankton
C) Squid
D) Seaweed

Question 18: Which animal is known for having cube-shaped poop?

A) Wombat
B) Koala
C) Kangaroo
D) Tasmanian Devil

Question 19: Which mammal has a heart that is about the size of a small car?

A) Elephant
B) Blue Whale
C) Rhinoceros
D) Hippopotamus

Question 20: Which fish is known for its ability to "walk" on land?

A) Clownfish
B) Archerfish
C) Mudskipper
D) Goby

Introduction to the Human Body

Welcome to the incredible world of the human body! Our bodies are marvels of nature, composed of intricate systems and structures that work together to sustain life. In this section, we will explore fascinating facts about the human body, including its skeletal, muscular, nervous, circulatory, respiratory, digestive, and other systems. Get ready to be amazed by the wonders of your own body!

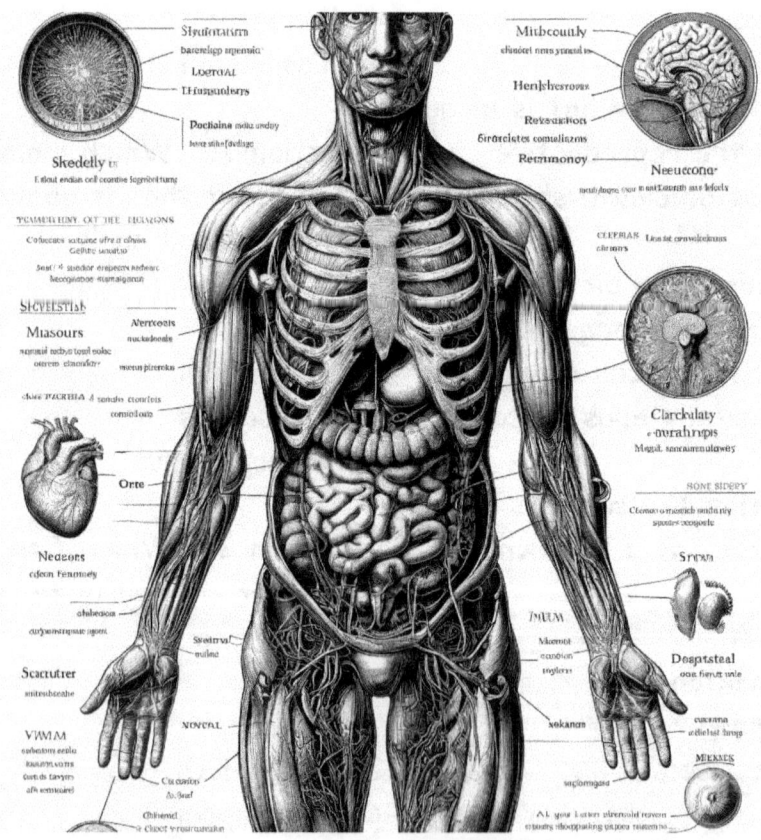

- **Fact 1**: The human skeleton is made up of 206 bones.

    "These bones provide structure, protect internal organs, and facilitate movement."

- **Fact 2**: Babies are born with approximately 270 bones.

  "Some of these bones fuse together as they grow, resulting in 206 bones in adulthood."

- **Fact 3**: The smallest bone in the human body is the stapes, located in the middle ear.

  "This tiny bone, also known as the stirrup, is crucial for hearing."

- **Fact 4**: The largest bone in the human body is the femur, or thigh bone.

  "The femur supports the weight of the body and allows for movement of the legs."

- **Fact 5**: There are 206 named bones in the human body.

  "These include long bones, short bones, flat bones, and irregular bones, each serving specific functions."

- **Fact 6**: The human hand has 27 bones.

  "These bones include the carpals, metacarpals, and phalanges, allowing for a wide range of movements and dexterity."

- **Fact 7**: The skull is composed of 22 bones.

  "These bones protect the brain and form the structure of the face.

- **Fact 8**: The spine consists of 33 vertebrae.

  "The vertebrae are divided into cervical, thoracic, lumbar, sacral, and coccygeal regions, providing support and flexibility."

- **Fact 9**: The hyoid bone is the only bone not connected to another bone.

  "Located in the neck, the hyoid bone supports the tongue and is involved in swallowing."

- **Fact 10**: The human body has over 600 muscles.

  "These muscles are responsible for movement, posture, and generating heat."

- **Fact 11**: The strongest muscle based on its size is the masseter, or jaw muscle.

  "The masseter can exert a force of up to 200 pounds on the molars."

- **Fact 12**: The gluteus maximus is the largest muscle in the human body.

  "This muscle in the buttocks is responsible for the movement of the hip and thigh."

- **Fact 13**: The smallest muscle in the body is the stapedius, located in the middle ear.

  "This muscle stabilizes the stapes bone and helps control the amplitude of sound waves.

- **Fact 14**: Humans have three types of muscle tissue: skeletal, smooth, and cardiac.

  "Skeletal muscles are voluntary and attached to bones, smooth muscles control involuntary movements, and cardiac muscle is found only in the heart."

- **Fact 15**: Muscle makes up about 40% of total body weight.

  "Muscles play a crucial role in movement, stability, and overall metabolism."

- **Fact 16**: The tongue is made up of eight muscles.

  "These muscles work together to allow for speech, swallowing, and taste."

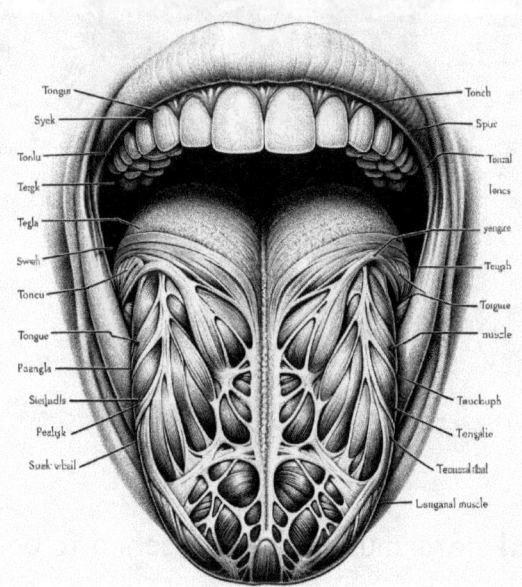

- **Fact 17**: Muscles work in pairs to move bones.

  "While one muscle contracts, its partner muscle relaxes, creating smooth and coordinated movements."

- **Fact 18**: The longest muscle in the body is the sartorius.

    "This muscle runs down the length of the thigh, from the hip to the knee, and is involved in flexing and rotating the leg."

- **Fact 19**: The heart is a muscle that pumps blood throughout the body.

    "It contracts about 100,000 times a day, pumping around 2,000 gallons of blood."

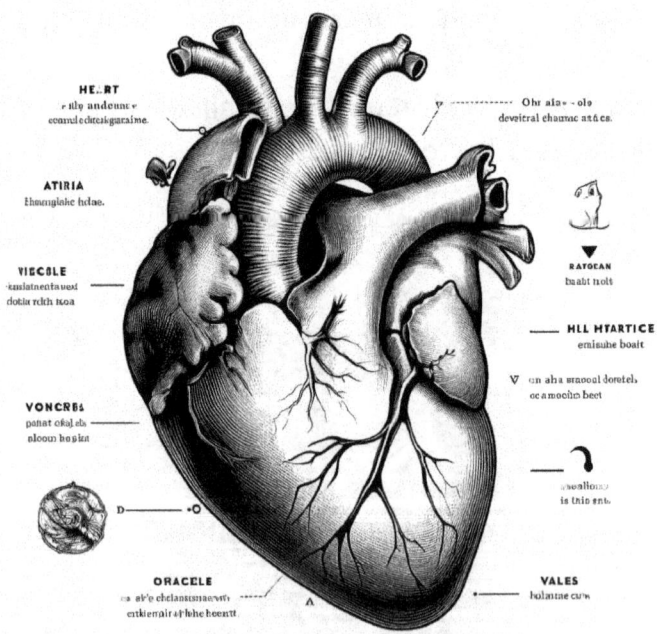

- **Fact 20**: Skeletal muscles are attached to bones by tendons.

    "Tendons are strong, fibrous connective tissues that transmit the force of muscle contractions to bones."

- **Fact 21**: The human brain contains approximately 86 billion neurons.

    "Neurons are specialized cells that transmit information through electrical and chemical signals."

- **Fact 22**: The brain accounts for about 2% of total body weight.

    "Despite its small size, the brain uses around 20% of the body's energy."

- **Fact 23**: The spinal cord is the main pathway for information connecting the brain to the rest of the body.

    "It transmits sensory and motor information and is protected by the vertebral column."

- **Fact 24**: The human brain can generate about 23 watts of power.

    "This amount of power is enough to light a small lightbulb."

- **Fact 25**: The nervous system is divided into the central nervous system (CNS) and the peripheral nervous system (PNS).

    "The CNS includes the brain and spinal cord, while the PNS consists of all other neural elements."

- **Fact 26**: The brain has four main lobes: frontal, parietal, temporal, and occipital.

    "Each lobe is responsible for different functions, such as movement, sensation, memory, and vision."

- **Fact 27**: The cerebellum is responsible for coordinating voluntary movements.

    "Located at the back of the brain, the cerebellum helps with balance, posture, and motor learning."

- **Fact 28**: The brainstem controls basic life functions.

    "These functions include breathing, heart rate, and blood pressure."

- **Fact 29**: The human brain has a left and right hemisphere.

    "Each hemisphere controls the opposite side of the body and is specialized for different functions."

- **Fact 30**: The human body has 12 pairs of cranial nerves.

    "These nerves emerge directly from the brain and control sensory and motor functions of the head and neck."

- **Fact 31**: The optic nerve transmits visual information from the retina to the brain.

    "This nerve is crucial for vision and is one of the 12 cranial nerves."

- **Fact 32**: Reflex actions are automatic responses to stimuli.

    "Reflexes are controlled by the spinal cord and occur without conscious thought."

- **Fact 33**: The autonomic nervous system regulates involuntary functions.

    "These functions include heart rate, digestion, and respiratory rate."

- **Fact 34**: Myelin is a fatty substance that insulates nerve fibers.

    "Myelin increases the speed of nerve signal transmission."

- **Fact 35**: The longest nerve in the body is the sciatic nerve.

    "It runs from the lower back down the legs and is responsible for motor and sensory functions of the lower extremities."

- **Fact 36**: The human brain can process information as quickly as 120 meters per second.

    "This rapid processing speed allows for quick reflexes and responses."

- **Fact 37**: The hippocampus is essential for memory formation.

  "Located in the temporal lobe, the hippocampus plays a key role in converting short-term memories into long-term memories."

- **Fact 38**: The hypothalamus regulates homeostasis.

  "This small brain region controls body temperature, hunger, thirst, and other essential functions."

- **Fact 39**: The amygdala is involved in emotional processing.

  "This almond-shaped structure in the brain is crucial for recognizing and reacting to emotional stimuli."

- **Fact 40**: Neurons communicate through synapses.

  "Synapses are junctions where neurons transmit signals to each other or to other cell types."

- **Fact 41**: The human heart beats about 100,000 times a day.

    "This continuous beating pumps around 2,000 gallons of blood throughout the body each day."

- **Fact 42**: Blood is pumped through a network of about 60,000 miles of blood vessels.

    "This extensive network includes arteries, veins, and capillaries that transport blood to every part of the body."

- **Fact 43**: The average adult has about 5 liters (1.3 gallons) of blood.

    "Blood is composed of red blood cells, white blood cells, platelets, and plasma."

- **Fact 44**: Red blood cells carry oxygen to the body's tissues.

    "These cells contain hemoglobin, a protein that binds to oxygen in the lungs and releases it to tissues."

- **Fact 45**: The heart has four chambers: two atria and two ventricles.

    "The atria receive blood, while the ventricles pump blood out to the lungs and the rest of the body."

- **Fact 46**: The aorta is the largest artery in the body.

    "It carries oxygenated blood from the heart to the rest of the body."

- **Fact 47**: Capillaries are the smallest blood vessels in the body.

    "They connect arteries and veins and facilitate the exchange of oxygen, carbon dioxide, and other substances between blood and tissues."

- **Fact 48**: White blood cells are part of the immune system.

    "These cells help defend the body against infections and foreign invaders."

- **Fact 49**: Platelets are responsible for blood clotting.

    "When a blood vessel is injured, platelets gather at the site and form a clot to stop bleeding."

- **Fact 50**: The average heart rate for an adult is about 72 beats per minute.

    "Heart rate can vary depending on factors like age, fitness level, and activity."

- **Fact 51**: The pulmonary artery carries deoxygenated blood from the heart to the lungs.

    "This is unique because most arteries carry oxygenated blood, but the pulmonary artery is an exception."

- **Fact 52**: The vena cava is the largest vein in the body.

    "It carries deoxygenated blood from the body back to the heart."

- **Fact 53**: The human body can produce up to 2 million red blood cells per second.

    "These cells are produced in the bone marrow and have a lifespan of about 120 days."

- **Fact 54**: The heart is located slightly to the left of the center of the chest.

    "It sits in the thoracic cavity, protected by the ribcage."

- **Fact 55**: Blood makes up about 7-8% of a person's body weight.

    "This percentage can vary slightly depending on an individual's size and composition."

- **Fact 56**: The coronary arteries supply blood to the heart muscle itself.

    "These arteries branch off from the aorta and ensure the heart gets the oxygen and nutrients it needs to function."

- **Fact 57**: Hemoglobin is the protein in red blood cells that carries oxygen.

  "It can bind to four oxygen molecules at a time, facilitating efficient oxygen transport."

- **Fact 58**: The heart can continue to beat even when disconnected from the body.

  "This is due to the heart's own electrical system, which can generate impulses independently of the brain."

- **Fact 59**: Blood pressure is the force exerted by circulating blood on the walls of blood vessels.

  "Normal blood pressure is typically around 120/80 mmHg, though it can vary based on age, health, and activity level."

- **Fact 60**: The circulatory system helps regulate body temperature.

  "Blood vessels can dilate to release heat or constrict to retain heat, helping maintain a stable body temperature."

- **Fact 61**: The human lungs contain about 300 million alveoli.

  "Alveoli are tiny air sacs where gas exchange occurs, allowing oxygen to enter the blood and carbon dioxide to be expelled."

- **Fact 62**: The surface area of the lungs is roughly the size of a tennis court.

  " This large surface area facilitates efficient gas exchange to meet the body's oxygen needs."

- **Fact 63**: The diaphragm is the primary muscle used in breathing.

  "This dome-shaped muscle contracts to expand the lungs during inhalation and relaxes during exhalation."

- **Fact 64**: The respiratory system is divided into the upper and lower respiratory tracts.

  "The upper tract includes the nose, nasal cavity, and pharynx, while the lower tract includes the larynx, trachea, bronchi, and lungs."

- **Fact 65**: The trachea, or windpipe, is about 4-5 inches long.

  "It connects the larynx to the bronchi and serves as the main airway to the lungs."

- **Fact 66**: Cilia are tiny hair-like structures in the respiratory tract.

  "They help trap and move particles out of the airways to keep the lungs clean."

- **Fact 67**: The right lung is larger than the left lung.

  "The left lung is smaller to make room for the heart."

- **Fact 68**: Oxygen makes up about 21% of the air we breathe.

  "The rest is mostly nitrogen, with trace amounts of other gases."

- **Fact 69**: The respiratory rate for an adult at rest is typically 12-20 breaths per minute.

   "This rate can increase with activity or stress."

- **Fact 70**: The lungs can hold about 6 liters of air at maximum capacity.

   "This volume is known as the total lung capacity and varies with age, sex, and physical condition."

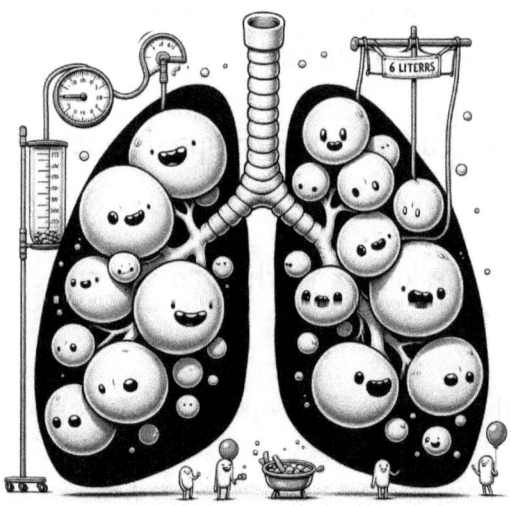

- **Fact 71**: The epiglottis prevents food from entering the trachea.

   "This flap of tissue closes over the trachea during swallowing, directing food and liquid into the esophagus."

- **Fact 72**: The respiratory system also plays a role in regulating blood pH.

   "By adjusting the rate of carbon dioxide removal, the respiratory system helps maintain the acid-base balance in the blood."

- **Fact 73**: The average person breathes in about 11,000 liters of air each day.

  "This constant exchange of air provides the oxygen needed for cellular respiration."

- **Fact 74**: The vocal cords are located in the larynx.

  "They vibrate to produce sound when air is expelled from the lungs."

- **Fact 75**: The process of breathing is called ventilation.

  "It involves both inhalation and exhalation to move air in and out of the lungs."

- **Fact 76**: The human digestive tract is about 30 feet long.

  "It includes the mouth, esophagus, stomach, small intestine, large intestine, rectum, and anus."

- **Fact 77**: The stomach produces hydrochloric acid to aid digestion.

  "This acid breaks down food and kills harmful bacteria."

- **Fact 78**: The small intestine is where most nutrient absorption occurs.

  "It is about 20 feet long and lined with villi and microvilli to increase surface area for absorption."

- **Fact 79**: The liver is the largest internal organ.

  "It performs over 500 functions, including detoxification, protein synthesis, and bile production."

- **Fact 80**: The pancreas produces enzymes and hormones.

  "It releases digestive enzymes into the small intestine and insulin and glucagon into the bloodstream to regulate blood sugar."

- **Fact 81**: The large intestine absorbs water and forms feces.

  "It is about 5 feet long and includes the cecum, colon, rectum, and anus."

- **Fact 82**: Saliva contains enzymes that begin the digestion of carbohydrates.

  "Amylase in saliva breaks down starches into simpler sugars."

- **Fact 83**: The esophagus is a muscular tube that connects the mouth to the stomach.

  "It uses peristalsis, a series of muscle contractions, to move food to the stomach."

- **Fact 84**: The stomach has three layers of muscle.

  "These layers churn food to mix it with digestive juices, forming a substance called chyme."

- **Fact 85**: The gallbladder stores and concentrates bile.

  "Bile, produced by the liver, helps digest fats in the small intestine."

- **Fact 86**: The digestive system is controlled by both the nervous and endocrine systems.

  "Hormones and nerve signals regulate the functions of the digestive organs."

- **Fact 87**: The appendix is a small, tube-like structure attached to the large intestine.

  "Its exact function is not well understood, but it is believed to play a role in gut immunity."

- **Fact 88**: The average time for food to travel through the digestive tract is about 24 to 72 hours.

  "This time can vary based on the type of food and individual digestive health."

- **Fact 89**: The inner lining of the stomach is replaced every few days.

    "This rapid turnover helps protect the stomach from being digested by its own acid."

- **Fact 90**: The mouth is the beginning of the digestive tract.

    "Digestion starts here with chewing and the action of saliva."

- **Fact 91**: The human eye can distinguish about 10 million different colors.

    "This incredible ability is due to the presence of cone cells in the retina that detect different wavelengths of light."

- **Fact 92**: The ear is responsible for both hearing and balance.

    "The inner ear contains the cochlea for hearing and the vestibular system for balance."

- **Fact 93**: The human nose can detect over 1 trillion different scents.

  "Olfactory receptors in the nose send signals to the brain, allowing us to perceive a vast array of smells."

- **Fact 94**: Taste buds are located on the tongue, soft palate, and throat.

  "These sensory organs detect sweet, sour, salty, bitter, and umami flavors."

- **Fact 95**: The average human has about 5 million scent receptors in the nose.

  "These receptors are highly sensitive and can detect even very low concentrations of odor molecules."

- **Fact 96**: The retina contains two types of photoreceptor cells: rods and cones.

  "Rods are responsible for vision in low light, while cones detect color and detail in bright light."

- **Fact 97**: The sense of touch is mediated by receptors in the skin.

  "These receptors can detect pressure, temperature, and pain, providing crucial information about our environment."

- **Fact 98**: The human tongue has about 10,000 taste buds.

  "Taste buds are replaced approximately every 10 to 14 days."

- **Fact 99**: The smallest muscles in the body are found in the middle ear.

  "These muscles, including the stapedius and tensor tympani, help protect the ear from loud noises."

- **Fact 100**: The human eye can process about 36,000 bits of information every hour.

  "This high processing speed allows for rapid visual perception and response."

- **Fact 101**: The immune system defends the body against harmful invaders.

  "It includes white blood cells, antibodies, and other components that work together to protect against infections and diseases."

- **Fact 102**: There are two main types of immunity: innate and adaptive.

  "Innate immunity is the body's first line of defense, while adaptive immunity develops over time and provides long-term protection."

- **Fact 103**: The spleen is an important organ in the immune system.

  "It filters blood, recycles old red blood cells, and helps fight certain bacteria."

- **Fact 104**: The thymus is where T cells mature.

    "T cells are a type of white blood cell that plays a critical role in the immune response."

- **Fact 105**: Bone marrow produces new blood cells, including immune cells.

    "It is the primary site of new blood cell production, including red blood cells, white blood cells, and platelets."

- **Fact 106**: Antibodies are proteins that recognize and neutralize foreign substances.

    "They are produced by B cells and play a key role in the body's defense against infections."

- **Fact 107**: Lymph nodes filter lymphatic fluid and trap pathogens.

    "They contain immune cells that can respond to infections by producing antibodies and destroying harmful invaders."

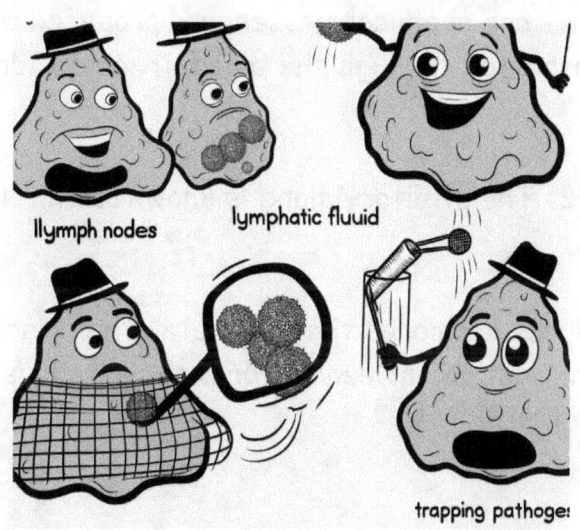

llymph nodes   lymphatic fluuid

trapping pathoge:

- **Fact 108**: The skin acts as a physical barrier against pathogens.

    "It is the body's largest organ and provides the first line of defense against infections."

- **Fact 109**: Fever is a common immune response to infection.

    "Raising the body's temperature helps inhibit the growth of pathogens and boosts the effectiveness of immune cells."

- **Fact 110**: The immune system can remember past infections.

    "Memory cells in the adaptive immune system provide long-term protection by recognizing and responding more rapidly to previously encountered pathogens."

- **Fact 111**: The endocrine system regulates bodily functions through hormones.

    "Hormones are chemical messengers produced by glands and transported through the bloodstream to target organs."

- **Fact 112**: The pituitary gland is known as the "master gland."

    "It produces hormones that regulate other glands in the endocrine system, influencing growth, metabolism, and reproduction."

- **Fact 113**: The thyroid gland regulates metabolism.

  "It produces hormones that control the rate at which the body uses energy and affects nearly every organ system."

- **Fact 114**: The pancreas has both endocrine and exocrine functions.

  "It produces insulin and glucagon to regulate blood sugar levels and digestive enzymes to break down food."

- **Fact 115**: The adrenal glands produce adrenaline.

  "Adrenaline, also known as epinephrine, prepares the body for "fight or flight" responses during stressful situations."

- **Fact 116**: The pineal gland produces melatonin.

  "Melatonin regulates sleep-wake cycles and is influenced by light exposure."

- **Fact 117**: The parathyroid glands regulate calcium levels in the blood.

  "These small glands produce parathyroid hormone, which helps maintain stable levels of calcium and phosphorus."

- **Fact 118**: The ovaries produce estrogen and progesterone.

  "These hormones regulate the female reproductive system and secondary sexual characteristics."

- **Fact 119**: The testes produce testosterone.

  "Testosterone regulates male reproductive functions and secondary sexual characteristics."

- **Fact 120**: Hormones can affect mood and behavior.

  "Hormonal changes can influence emotions, stress levels, and overall mental health, demonstrating the interconnectedness of the endocrine and nervous systems."

- **Fact 121**: The human nose can remember 50,000 different scents.

  "Our sense of smell is incredibly powerful and capable of distinguishing a vast array of scents, often triggering vivid memories and emotions."

Human Body Trivia Questions-1
(Answers at the end)

Question 1: How many bones are in the adult human body?
A) 206
B) 270
C) 195
D) 220

Question 2: Which is the largest organ in the human body?
A) Liver
B) Heart
C) Skin
D) Lungs

Question 3: What is the smallest bone in the human body?
A) Femur
B) Stapes
C) Ulna
D) Tibia

Question 4: How many chambers are there in the human heart?
A) 2
B) 4
C) 3
D) 5

Question 5: What is the primary function of red blood cells?
A) Fighting infections
B) Clotting blood
C) Carrying oxygen
D) Producing hormones

Question 6: The human brain is part of which system?
A) Respiratory System
B) Digestive System
C) Nervous System
D) Endocrine System

Question 7: Which muscle is known as the strongest based on its size?
A) Biceps
B) Gluteus Maximus
C) Sartorius
D) Masseter

Question 8: Where are the alveoli located?
A) Heart
B) Brain
C) Lungs
D) Liver

Human Body Trivia Questions-2
(Answers at the end)

Question 9: How many lobes are in the human brain?

A) 2
B) 4
C) 6
D) 8

Question 10: What does the pituitary gland regulate?

A) Digestion
B) Metabolism
C) Heart Rate
D) Hormone Production

Question 11: Which part of the eye detects light and color?

A) Lens
B) Cornea
C) Retina
D) Iris

Question 12: What type of joint is the shoulder?

A) Hinge
B) Ball and Socket
C) Pivot
D) Saddle

Question 13: Which organ is responsible for filtering blood?

A) Heart
B) Lungs
C) Liver
D) Kidneys

Question 14: What is the main function of the large intestine?

A) Absorb nutrients
B) Produce enzymes
C) Absorb water and form feces
D) Store bile

Human Body Trivia Questions-3
(Answers at the end)

Question 15: How many taste buds does the average human tongue have?

A) 5,000
B) 10,000
C) 15,000
D) 20,000

Question 16: Which hormone is primarily involved in regulating sleep?

A) Insulin
B) Adrenaline
C) Melatonin
D) Cortisol

Question 17: What part of the cell is known as the "powerhouse"?

A) Nucleus
B) Ribosome
C) Mitochondrion
D) Golgi Apparatus

Question 18: What substance in the blood helps in clotting?

A) Red blood cells
B) White blood cells
C) Platelets
D) Plasma

Question 19: Which part of the nervous system controls involuntary actions?

A) Central Nervous System
B) Peripheral Nervous System
C) Autonomic Nervous System
D) Somatic Nervous System

Question 20: What is the largest artery in the human body?

A) Pulmonary artery
B) Femoral artery
C) Carotid artery
D) Aorta

Introduction to Technology

Technology has revolutionized our world, transforming the way we live, work, and communicate. From the invention of the wheel to the latest advancements in artificial intelligence, technology continues to drive innovation and progress. In this section, we will explore fascinating facts about various aspects of technology, including historical milestones, modern gadgets, computing, the internet, and future trends. Get ready to dive into the world of technology and discover some amazing insights!

- **Fact 1**: The first computer programmer was Ada Lovelace.

    "In the 1840s, Ada Lovelace wrote the first algorithm intended to be processed by Charles Babbage's early mechanical general-purpose computer, the Analytical Engine.."

- **Fact 2**: The term "bug" to describe a software glitch originated with a real bug.

  "In 1947, Grace Hopper found a moth causing a malfunction in the Mark II computer, and the term "debugging" was born."

- **Fact 3**: The first successful personal computer was the Apple II.

  "Released in 1977, the Apple II was a major success and helped launch the personal computing revolution."

- **Fact 4**: The World Wide Web was invented by Tim Berners-Lee.

  "In 1989, Tim Berners-Lee created the World Wide Web while working at CERN, revolutionizing how information is shared and accessed."

- **Fact 5**: The first smartphone was the IBM Simon Personal Communicator.

  "Released in 1994, the IBM Simon was the first device to combine a mobile phone with computing capabilities, including email and fax."

- **Fact 6**: The first video game console was the Magnavox Odyssey.

  "Released in 1972, the Magnavox Odyssey was the first commercial home video game console, paving the way for future gaming systems."

- **Fact 7**: The first computer virus was called "Creeper."

  "In 1971, the Creeper virus was created as an experimental self-replicating program on the ARPANET, the precursor to the internet."

- **Fact 8**: The first 1 GB hard drive was released in 1980.

  "IBM introduced the first gigabyte hard drive, which weighed over 500 pounds and cost $40,000."

- **Fact 9**: The first email was sent by Ray Tomlinson in 1971.

  "Ray Tomlinson sent the first email using the ARPANET, the predecessor to the internet, and it was a simple test message."

- **Fact 10**: The first website is still online.

  "Tim Berners-Lee created the first website, which is still accessible today at http://info.cern.ch"

- **Fact 11**: The first iPhone was released in 2007.

  "Apple revolutionized the smartphone market with the release of the first iPhone, introducing a touch-screen interface and a wide range of applications."

- **Fact 12**: The average smartphone has more computing power than the Apollo 11 mission's computers.

  "Modern smartphones are incredibly powerful, surpassing the computational capabilities of the computers used in the Apollo 11 moon landing.

- **Fact 13**: The world's largest digital camera is being built for the Vera C. Rubin Observatory.

   "The camera will have a resolution of 3.2 gigapixels and will be used to survey the night sky, capturing detailed images of the universe."

- **Fact 14**: The Amazon Echo was the first widely adopted smart speaker.

   "Released in 2014, the Amazon Echo, powered by the Alexa voice assistant, popularized the use of smart speakers in homes."

- **Fact 15**: The first 3D-printed organ was a human bladder.

   "In 1999, scientists successfully 3D-printed a human bladder and implanted it in a patient, marking a significant advancement in medical technology."

- **Fact 16**: The Google Glass was an early attempt at augmented reality glasses.

    "Released in 2013, Google Glass featured a head-mounted display but faced criticism for privacy and usability issues, leading to its discontinuation for consumers."

- **Fact 17**: The PlayStation 2 is the best-selling video game console of all time.

    "Released in 2000, the PlayStation 2 sold over 155 million units worldwide, making it the most popular console ever."

- **Fact 18**: The first digital camera was created in 1975 by Kodak.

    "The prototype digital camera had a resolution of 0.01 megapixels and took 23 seconds to capture its first image."

- **Fact 19**: The first consumer drone was the Parrot AR Drone.

    "Released in 2010, the Parrot AR Drone was one of the first consumer-grade drones, controlled via smartphone or tablet."

- **Fact 20**: The first smartwatch was the Seiko Data 2000.

  "Released in 1983, the Seiko Data 2000 could store memos and had a keyboard for data entry, making it an early precursor to modern smartwatches."

- **Fact 21**: Moore's Law predicts the doubling of transistors on a microchip every two years.

  "This observation by Gordon Moore has held true for decades, leading to exponential growth in computing power."

- **Fact 22**: The first microprocessor was the Intel 4004.

  "Released in 1971, the Intel 4004 was the first commercially available microprocessor, revolutionizing computing technology."

- **Fact 23**: The term "bit" is short for "binary digit."

  "A bit is the smallest unit of data in computing, representing a 0 or 1 in binary code."

- **Fact 24**: Quantum computers use qubits instead of bits.

    "Qubits can represent both 0 and 1 simultaneously, allowing quantum computers to solve complex problems much faster than classical computers."

- **Fact 25**: The first graphical user interface (GUI) was developed by Xerox PARC.

    "Xerox's GUI, introduced in the 1970s, laid the foundation for modern operating systems like Windows and macOS."

- **Fact 26**: The world's fastest supercomputer is Fugaku.

    "As of 2021, Fugaku, developed by RIKEN and Fujitsu in Japan, is the world's fastest supercomputer, capable of over 442 petaflops."

- **Fact 27**: The first computer mouse was made of wood.

    "Invented by Douglas Engelbart in 1964, the original mouse was a wooden shell with two metal wheels."

- **Fact 28**: The ENIAC was the first general-purpose electronic digital computer.

    "Completed in 1945, the ENIAC (Electronic Numerical Integrator and Computer) was used for calculations in military and scientific applications."

- **Fact 29:** The first operating system was developed in the 1950s.

  "Early operating systems like GM-NAA I/O were created to manage hardware and software resources on mainframe computers."

- **Fact 30:** The Unicode standard allows computers to display and manipulate text from multiple languages.

  "Unicode provides a unique number for every character, enabling consistent encoding, representation, and handling of text across different platforms."

- **Fact 31:** The first website was created in 1991.

  "Tim Berners-Lee launched the first website, which explained the basics of the World Wide Web project."

- **Fact 32:** The term "spam" for unsolicited emails comes from a Monty Python sketch.

  "The sketch featured a diner where every item on the menu included spam, leading to the association with excessive and unwanted communication."

- **Fact 33:** Google processes over 3.5 billion searches per day.

  "This vast number of searches highlights the importance of search engines in modern information retrieval."

- **Fact 34:** The first domain name ever registered was symbolics.com

    "Registered on March 15, 1985, symbolics.com was the first domain name, belonging to a computer manufacturer."

- **Fact 35**: The majority of internet traffic is generated by video streaming.

    "Platforms like YouTube, Netflix, and other streaming services account for a significant portion of global internet traffic."

- **Fact 36:** The Internet of Things (IoT) refers to the network of connected devices.

    "IoT devices, such as smart home gadgets, wearables, and industrial sensors, communicate and exchange data over the internet."

- **Fact 37:** The first email spam was sent in 1978.

  "Gary Thuerk, a marketer, sent an unsolicited email to 400 users on the ARPANET, promoting a new product."

- **Fact 38:** The most popular social media platform is Facebook.

  "With over 2.8 billion monthly active users, Facebook is the largest social media network in the world."

- **Fact 39:** The term "Wi-Fi" stands for "Wireless Fidelity."

  "Wi-Fi is a technology that allows devices to connect to the internet wirelessly using radio waves."

- **Fact 40:** The first viral video on YouTube was "Me at the zoo."

  "Uploaded in 2005, this video features YouTube co-founder Jawed Karim at the San Diego Zoo and has since become a significant piece of internet history."

- **Fact 41:** Artificial intelligence (AI) is becoming increasingly integrated into everyday life.

  "AI technologies, such as machine learning and natural language processing, are used in various applications, including virtual assistants, autonomous vehicles, and personalized recommendations."

- **Fact 42:** Quantum computing has the potential to revolutionize problem-solving.

    "Quantum computers can process information much faster than classical computers, potentially solving complex problems in fields like cryptography, medicine, and material science."

- **Fact 43**: 5G technology offers significantly faster internet speeds.

    "The fifth generation of mobile network technology, 5G, provides faster data transfer rates, lower latency, and more reliable connections compared to previous generations."

- **Fact 44:** Augmented reality (AR) and virtual reality (VR) are expanding into new industries.

    "AR and VR technologies are being used in healthcare, education, entertainment, and retail to create immersive experiences and enhance training and visualization."

- **Fact 45:** Blockchain technology ensures secure and transparent transactions.

    "Originally developed for cryptocurrencies like Bitcoin, blockchain technology is now being used for secure data sharing, supply chain management, and digital identity verification."

- **Fact 46**: Autonomous vehicles are being developed for various modes of transportation.

  "Self-driving cars, trucks, and even drones are being designed to improve safety, efficiency, and convenience in transportation."

- **Fact 47**: Biometric authentication is becoming more common.

  "Technologies like fingerprint scanning, facial recognition, and iris scanning are being used for secure and convenient access to devices and services."

- **Fact 48**: The Internet of Things (IoT) is connecting more devices than ever before.

  "IoT technology allows everyday objects to communicate with each other and the internet, creating smart homes, cities, and industries."

- **Fact 49**: 3D printing is revolutionizing manufacturing.

  "3D printing, or additive manufacturing, allows for the creation of complex and customized objects, reducing waste and production time in industries like healthcare, aerospace, and automotive."

- **Fact 50**: Renewable energy technologies are advancing rapidly.

  "Innovations in solar, wind, and battery technologies are making renewable energy more efficient and accessible, contributing to a more sustainable future."

- **Fact 51:** The first industrial robot was introduced in 1961.

    "Unimate, the first industrial robot, was used in a General Motors assembly line to handle hot metal and perform repetitive tasks."

- **Fact 52**: Robots are being used in surgery to improve precision.

    "Surgical robots, such as the Da Vinci system, allow surgeons to perform complex procedures with greater accuracy and minimal invasiveness."

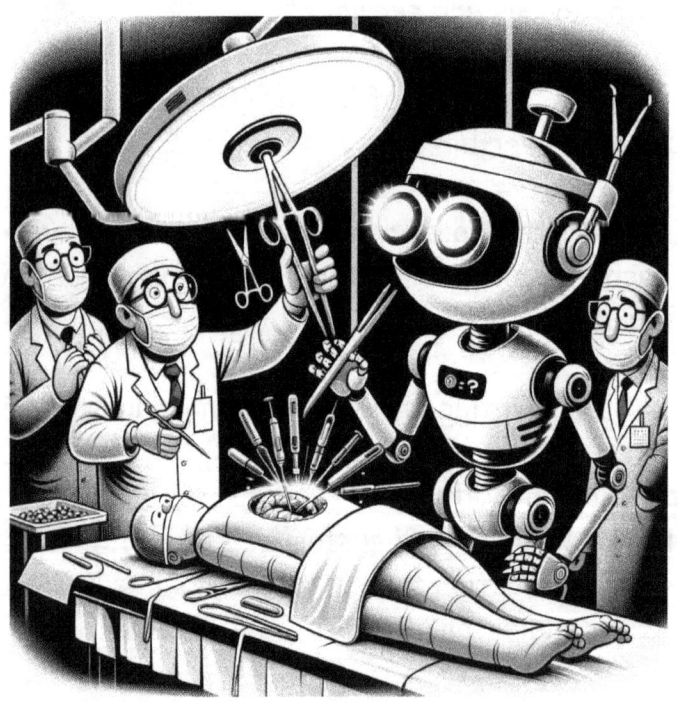

- **Fact 53:** Soft robotics is a growing field.

    "Soft robots are made from flexible materials, allowing them to mimic natural movements and interact safely with humans and delicate objects."

- **Fact 54**: Robots are exploring space.

   "Robotic rovers, like NASA's Curiosity and Perseverance, are exploring Mars, collecting data, and searching for signs of past life."

- **Fact 55**: Robots can learn through machine learning algorithms.

   "Machine learning allows robots to improve their performance by analyzing data and learning from their experiences."

- **Fact 56**: Exoskeletons assist people with mobility impairments.

   "Wearable robotic exoskeletons can help individuals with paralysis or mobility issues to walk and perform daily activities."

- **Fact 57**: Robots are being used in agriculture to increase efficiency.

   "Agricultural robots can perform tasks such as planting, harvesting, and monitoring crop health, reducing labor costs and increasing productivity"

- **Fact 58**: Drones are being used for aerial photography and mapping.

   "Unmanned aerial vehicles (UAVs) equipped with cameras and sensors are used for capturing images and creating detailed maps of large areas"

- **Fact 59**: Social robots interact with humans.

    "Robots like Pepper and Nao are designed to engage with people, providing companionship, assistance, and education."

- **Fact 60**: Autonomous underwater robots explore the ocean.

    "These robots can dive deep into the ocean to study marine life, map the seafloor, and monitor environmental changes."

- **Fact 61**: The first transatlantic telegraph cable was laid in 1858.

    "This cable connected North America and Europe, allowing for nearly instantaneous communication across the Atlantic Ocean."

- **Fact 62**: The first mobile phone call was made in 1973.

    "Martin Cooper of Motorola made the first handheld mobile phone call, marking the beginning of the mobile communication revolution."

- **Fact 63**: Fiber optic cables transmit data using light.

    "These cables use light signals to transmit data at high speeds over long distances, providing faster and more reliable internet connections"

- **Fact 64:** The first satellite, Sputnik 1, was launched in 1957.

    "Sputnik 1, launched by the Soviet Union, was the first artificial satellite, paving the way for satellite communication and space exploration."

- **Fact 65**: Bluetooth technology allows for short-range wireless communication.

    "Named after a Danish king, Bluetooth enables devices to connect and share data wirelessly over short distances."

- **Fact 66:** The first email was sent in 1971.

    "Ray Tomlinson sent the first email on the ARPANET, which was a simple test message."

- **Fact 67:** Morse code was one of the earliest forms of digital communication.

  "Invented by Samuel Morse, Morse code uses sequences of dots and dashes to represent letters and numbers, allowing for long-distance communication via telegraph."

- **Fact 68:** The QR code was invented in 1994 by Denso Wave.

  "QR codes are two-dimensional barcodes that can store information and are easily scanned by smartphones and other devices."

- **Fact 69:** Voice over Internet Protocol (VoIP) allows for voice communication over the internet.

  "VoIP technology, used in services like Skype and Zoom, converts voice signals into digital data and transmits them over the internet."

- **Fact 70:** The first webcam was used to monitor a coffee pot.

  "In 1991, researchers at the University of Cambridge set up a webcam to check the status of a coffee pot, creating one of the earliest examples of internet-connected cameras."

- **Fact 71**: The first high-level programming language was FORTRAN.

  "Developed in the 1950s by IBM, FORTRAN (short for "Formula Translation") was designed for scientific and engineering calculations."

- **Fact 72**: The first computer game was "Spacewar!.

  "Created in 1962 by MIT students, "Spacewar!" is considered one of the earliest video games and was played on a PDP-1 minicomputer."

- **Fact 73**: Open-source software allows anyone to view and modify its code.

  "Projects like Linux and Mozilla Firefox are open-source, enabling developers worldwide to contribute to their development and improvement."

- **Fact 74:** The Python programming language was named after Monty Python.

  "Guido van Rossum, the creator of Python, named the language after the British comedy group Monty Python, reflecting his desire for a fun and approachable programming language."

- **Fact 75:** The Java programming language was initially called Oak.

  "Java, created by Sun Microsystems, was originally named Oak after an oak tree outside the developer's office, but the name was changed due to trademark issues."

- **Fact 76:** CAPTCHA stands for "Completely Automated Public Turing test to tell Computers and Humans Apart.

    "CAPTCHAs are used to prevent automated bots from accessing websites by requiring users to complete a test that only humans can pass."

- **Fact 77:** The Linux operating system kernel was created by Linus Torvalds.

    "In 1991, Linus Torvalds released the Linux kernel, which has since become the foundation for numerous operating systems and is widely used in servers, desktops, and mobile devices."

- **Fact 78:** The term "debugging" was popularized by Grace Hopper.

    "Grace Hopper found and removed a moth that was causing a malfunction in the Mark II computer, coining the term 'debugging.'"

- **Fact 79:** Git is a version control system created by Linus Torvalds.

    "Git, developed in 2005, is widely used by developers to track changes in source code during software development."

- **Fact 80:** The Turing Award is often considered the "Nobel Prize of Computing.

    "Named after Alan Turing, the Turing Award is given annually by the Association for Computing Machinery (ACM) to individuals who have made significant contributions to the field of computer science."

- **Fact 81:** The first known computer virus was the "Creeper" virus.

    "Created in 1971, the Creeper virus infected ARPANET computers and displayed the message, 'I'm the creeper: catch me if you can.'"

- **Fact 82:** Two-factor authentication (2FA) adds an extra layer of security.

    "2FA requires users to provide two forms of identification, such as a password and a verification code, to access an account, enhancing security."

- **Fact 83:** Phishing is a common cyberattack method.

    "Phishing involves tricking individuals into providing sensitive information, such as passwords and credit card numbers, by pretending to be a trustworthy entity."

- **Fact 84:** Ransomware encrypts files and demands payment for their release.

    "Ransomware attacks involve encrypting a victim's files and demanding a ransom, often in cryptocurrency, to decrypt them."

- **Fact 85:** The Stuxnet worm targeted industrial control systems.

    "Discovered in 2010, Stuxnet was a sophisticated computer worm that targeted Iran's nuclear facilities, disrupting their operations."

- **Fact 86:** A firewall helps protect networks from unauthorized access.

    "Firewalls monitor and control incoming and outgoing network traffic based on predetermined security rules"

- **Fact 87**: Encryption is the process of encoding data to prevent unauthorized access.

    "Encrypted data can only be read by someone with the correct decryption key, ensuring the confidentiality and integrity of sensitive information."

- **Fact 88**: The General Data Protection Regulation (GDPR) protects personal data in the EU.

    "Enacted in 2018, the GDPR sets guidelines for the collection and processing of personal data, enhancing privacy and data protection for individuals in the European Union."

- **Fact 89**: The first antivirus software was created in 1987.

    "Developed by Andreas Lüning and Kai Figge, the first antivirus software was designed to detect and remove computer viruses."

- **Fact 90**: White hat hackers help improve cybersecurity.

    "Ethical hackers, or white hat hackers, use their skills to identify and fix security vulnerabilities, helping organizations protect against cyber threats."

- **Fact 91**: The first video game was "Tennis for Two.".

    "Created in 1958 by physicist William Higinbotham, "Tennis for Two" was an early interactive game displayed on an oscilloscope."

- **Fact 92:** The highest-grossing video game franchise is "Pokémon."

    "With billions in revenue from games, merchandise, and media, Pokémon is the most successful video game franchise of all time."

- **Fact 93**: The Nintendo Game Boy was the first handheld gaming console to achieve widespread success.

    "Released in 1989, the Game Boy sold over 118 million units and popularized portable gaming."

- **Fact 94:** The first home video game console was the Magnavox Odyssey.

    "Released in 1972, the Magnavox Odyssey was the first commercial home video game console, paving the way for future gaming systems."

- **Fact 95:** "Minecraft" is the best-selling video game of all time.

    "As of 2021, "Minecraft" has sold over 200 million copies worldwide, making it the best-selling video game ever."

- **Fact 96**: The "Easter egg" in video games refers to hidden messages or features.

    "The term originated with the game "Adventure" for the Atari 2600, which included a hidden message from the programmer."

- **Fact 97**: The first 3D video game was "3D Monster Maze."

    "Released in 1981, "3D Monster Maze" was a pioneering 3D game for the Sinclair ZX81 home computer."

- **Fact 98**: The first commercially successful video game was "Pong."

    "Developed by Atari and released in 1972, "Pong" was a simple table tennis game that became a cultural phenomenon."

- **Fact 99**: The longest video game marathon lasted over 35 hours.

    "In 2015, a team of gamers set a world record for the longest video game marathon, playing "Destiny" for over 35 hours straight."

- **Fact 100**: Neural interfaces allow direct communication between the brain and external devices..

    "Technologies like Brain-Computer Interfaces (BCIs) enable people to control devices using their thoughts, with potential applications in healthcare and communication."

- **Fact 101**: Edge computing processes data closer to its source.

    "By processing data near the source of generation, edge computing reduces latency and bandwidth usage, improving the performance of applications like IoT devices."

- **Fact 102**: Holographic displays create 3D images without the need for special glasses.

    "Holographic technology projects images in three dimensions, allowing viewers to see and interact with 3D visuals from different angles."

- **Fact 103**: Smart cities use technology to improve urban living..

    "Smart cities integrate IoT, AI, and data analytics to enhance infrastructure, reduce energy consumption, and improve the quality of life for residents."

- **Fact 104**: Wearable technology monitors health and fitness.

    "Devices like smartwatches and fitness trackers collect data on physical activity, heart rate, sleep patterns, and more, helping users maintain a healthy lifestyle."

- **Fact 105**: Digital twins create virtual replicas of physical objects.

    "Digital twins simulate real-world objects and systems, allowing for testing, monitoring, and optimization without physical prototypes."

- **Fact 106**: Genetic engineering can modify organisms at the DNA level.

  "Techniques like CRISPR-Cas9 enable precise editing of genetic material, with potential applications in medicine, agriculture, and biotechnology."

- **Fact 107**: Autonomous robots can perform tasks without human intervention.

  "These robots use AI and machine learning to navigate environments, make decisions, and complete tasks independently, from manufacturing to delivery services."

- **Fact 108**: Telemedicine allows healthcare providers to consult with patients remotely.

  "Through video calls and online platforms, telemedicine provides convenient access to medical care, especially for those in remote or underserved areas."

- **Fact 109**: CRISPR technology enables precise genetic editing..

  "CRISPR-Cas9 is a revolutionary tool for editing genes, allowing scientists to modify DNA sequences and potentially cure genetic disorders."

- **Fact 110**: 3D printing is being used to create custom prosthetics.

  "3D printing technology allows for the production of personalized prosthetic limbs that fit patients perfectly, improving comfort and functionality."

- **Fact 111**: Nanotechnology is being used to develop targeted drug delivery systems.

    "Nanoparticles can deliver medications directly to specific cells or tissues, increasing the effectiveness of treatments and reducing side effects."

- **Fact 112**: AI algorithms are being used to diagnose diseases.

    "Machine learning algorithms can analyze medical images and data to identify patterns and assist in diagnosing conditions like cancer and heart disease."

- **Fact 113**: Virtual reality (VR) is being used for pain management.

    "VR can distract patients from pain during medical procedures or rehabilitation, providing a non-pharmacological method of pain relief."

- **Fact 114:** Gene therapy has the potential to cure genetic disorders.

    "Gene therapy involves inserting, altering, or removing genes within an individual's cells to treat or prevent disease, offering hope for conditions previously considered untreatable."

- **Fact 112**: AI can analyze large datasets to identify patterns and make predictions.

    "AI algorithms excel at processing and analyzing vast amounts of data, uncovering insights that can inform decision-making and forecasting."

- **Fact 113:** AI-powered chatbots can handle customer service inquiries.

    "Businesses use AI chatbots to provide 24/7 customer support, answering common questions and resolving issues efficiently."

- **Fact 114:** AI can enhance cybersecurity by detecting anomalies.

  "AI systems monitor network activity and identify unusual patterns that may indicate a cyber attack, allowing for quicker response times."

- **Fact 115:** AI is transforming healthcare with predictive analytics.

  "AI algorithms analyze patient data to predict disease outbreaks, identify high-risk patients, and recommend personalized treatments."

- **Fact 116:** The average time to identify a data breach is 207 days.

  "It often takes months for organizations to detect a data breach, during which time significant damage can occur."

- **Fact 117:** Cybersecurity frameworks like NIST provide guidelines for protecting information.

  "The National Institute of Standards and Technology (NIST) framework offers best practices for managing and reducing cybersecurity risks."

- **Fact 118:** Zero-day vulnerabilities are exploited before developers can issue a patch.

  "These are security flaws unknown to the software developer and can be used by attackers to gain unauthorized access."

- **Fact 119:** The Dark Web is a part of the internet not indexed by traditional search engines.

  "The Dark Web is often used for illicit activities, including the sale of stolen data and illegal goods."

- **Fact 120:** Ransomware-as-a-Service (RaaS) allows cybercriminals to lease ransomware tools.

  "RaaS platforms provide ready-made ransomware kits to less skilled hackers, increasing the prevalence of ransomware attacks."

- **Fact 121:** Endpoint detection and response (EDR) tools monitor and analyze activity on devices.

  "EDR solutions help detect and respond to cyber threats by collecting and analyzing data from endpoints like computers and mobile devices."

Technology Trivia Questions-1
(Answers at the end)

Question 1: Who is considered the first computer programmer?

A) Alan Turing
B) Grace Hopper
C) Ada Lovelace
D) Charles Babbage

Question 2: What was the first video game console ever released?

A) Atari 2600
B) Nintendo Entertainment System
C) Sega Genesis
D) Magnavox Odyssey

Question 3: What year was the World Wide Web invented?

A) 1989
B) 1995
C) 2000
D) 1985

Question 4: Which company released the first successful personal computer, the Apple II?

A) IBM
B) Microsoft
C) Apple
D) Commodore

Question 5: What does AI stand for?

A) Artificial Integration
B) Automated Intelligence
C) Artificial Intelligence
D) Automated Interaction

Question 6: What is the primary function of a firewall in a computer network?

A) Encrypting data
B) Managing passwords
C) Filtering network traffic
D) Scanning for viruses

Question 7: Which AI technology allows machines to learn from data?

A) Quantum Computing
B) Blockchain
C) Machine Learning
D) 3D Printing

Technology Trivia Questions-2
(Answers at the end)

Question 8: What type of cyberattack involves overwhelming a system with traffic to make it unavailable?
A) Phishing
B) Malware
C) Denial-of-Service (DoS)
D) Ransomware

Question 9: Which programming language was named after a British comedy group?
A) Java
B) Python
C) Ruby
D) Perl

Question 10: What was the first smartphone ever released?
A) BlackBerry
B) Nokia 9000
C) IBM Simon
D) iPhone

Question 11: Which AI application involves the creation of realistic fake videos?
A) Chatbots
B) Deepfakes
C) Predictive Analytics
D) Autonomous Vehicles

Question 12: What is the primary goal of ransomware?
A) To steal data
B) To delete files
C) To encrypt data and demand a ransom for its release
D) To spread to other devices

Question 13: What does the term "phishing" refer to in cybersecurity?
A) Overloading a system with traffic
B) Unauthorized data access
C) Tricking individuals into providing sensitive information
D) Installing malicious software

Technology Trivia Questions-3
(Answers at the end)

Question 14: What does CAPTCHA stand for?

A) Computer Automated Public Turing Test to Help Computers
B) Completely Automated Public Turing test to tell Computers and Humans Apart
C) Comprehensive Automated Protection Test for Human Access
D) Complete Automated Programming Test to Help Computers

Question 15: What was the first high-level programming language?

A) C++
B) Python
C) FORTRAN
D) Java

Question 16: What is the world's largest digital camera being built for?

A) Astronomy
B) Medicine
C) Military
D) Robotics

Question 17: Which cybersecurity measure involves setting up a decoy system to attract attackers?

A) Phishing
B) Honeypot
C) Firewall
D) VPN

Question 18: What technology uses qubits instead of bits for processing information?

A) Blockchain
B) Fiber Optics
C) Quantum Computing
D) Machine Learning

Technology Trivia Questions-4
(Answers at the end)

Question 19: What does 5G technology primarily improve in mobile networks?

A) Battery life
B) Internet speeds and latency
C) Device size
D) Screen resolution

Question 20: What is the primary function of the blockchain technology?

A) Enhanced graphics processing
B) Secure and transparent transactions
C) Faster internet speeds
D) Data storage

Question 21: Who invented the World Wide Web?

A) Bill Gates
B) Tim Berners-Lee
C) Steve Jobs
D) Mark Zuckerberg

Question 22: What was the first computer virus called?

A) Morris Worm
B) ILOVEYOU
C) Stuxnet
D) Creeper

Question 23: What is the main function of a neural network in AI?

A) Encrypting data
B) Managing network traffic
C) Simulating the human brain to recognize patterns
D) Running computer simulations

Question 24: Which cybersecurity practice involves regularly updating software to fix vulnerabilities?
A) Phishing
B) Patching
C) Firewalls
D) Encryption

Answers

Space Trivia Answers

Question 1 - C   Question 11 - B

Question 2 - B   Question 12 - A

Question 3 - A   Question 13 - B

Question 4 - D   Question 14 - B

Question 5 - B   Question 15 - B

Question 6 - C   Question 16 - D

Question 7 - B   Question 17 - C

Question 8 - D   Question 18 - B

Question 9 - C   Question 19 - D

Question 10 - B   Question 20 - B

Earth Trivia Answers

Question 1 - A   Question 11 - B

Question 2 - D   Question 12 - D

Question 3 - B   Question 13 - C

Question 4 - C   Question 14 - C

Question 5 - D   Question 15 - A

Question 6 - B   Question 16 - C

Question 7 - B   Question 17 - B

Question 8 - C   Question 18 - C

Question 9 - C   Question 19 - B

Question 10 - C   Question 20 - C

Animal Trivia Answers

Question 1 - C   Question 11 - D

Question 2 - A   Question 12 - C

Question 3 - A   Question 13 - B

Question 4 - A   Question 14 - B

Question 5 - B   Question 15 - C

Question 6 - A   Question 16 - C

Question 7 - A   Question 17 - B

Question 8 - B   Question 18 - A

Question 9 - C   Question 19 - B

Question 10 - B   Question 20 - C

Human Body Trivia Answers

Question 1 - A   Question 11 - C

Question 2 - C   Question 12 - B

Question 3 - B   Question 13 - D

Question 4 - B   Question 14 - C

Question 5 - C   Question 15 - B

Question 6 - C   Question 16 - C

Question 7 - D   Question 17 - C

Question 8 - C   Question 18 - C

Question 9 - B   Question 19 - C

Question 10 - D   Question 20 - D

Technology Trivia Answers

Question 1 - C   Question 13 - C
Question 2 - D   Question 14 - B
Question 3 - A   Question 15 - C
Question 4 - C   Question 16 - A
Question 5 - C   Question 17 - B
Question 6 - C   Question 18 - C
Question 7 - C   Question 19 - B
Question 8 - C   Question 20 - B
Question 9 - B   Question 21 - B
Question 10 - C   Question 22 - D
Question 11 - B   Question 23 - C
Question 12 - C   Question 24 - B

END

www.ingramcontent.com/pod-product-compliance
Lightning Source LLC
Chambersburg PA
CBHW050218230526
45470CB00001B/437